調理好體質

想要懷孕，就從吃對食物開始

Prepare your body for pregnancy:
Start with the right diet

作者簡介

　　芳姐於香港大學專業進修學院修畢「中醫全科文憑」及修讀中醫全科學士多個課程。自 1996 年起在各大報章及雜誌撰寫食療專欄，作品散見於新晚報、大公報、壹蘋果健康網、加拿大明報周刊「樂在明廚」及北美洲「品」美食時尚雜誌。

　　芳姐曾任「中華廚藝學院」健康美食班客座講師，現任僱員再培訓局中心、職訓局中心的陪月班導師及「靚湯工房」榮譽食療顧問。

　　為推廣食療心得，芳姐自 2007 年開始在網上撰寫食療雜誌「芳姐保健湯餸」，深獲網民支持和愛戴。近年開始將其心得著作成書，著有《每天一杯養生茶》、《坐月天書》、《每天一杯養生茶 2》、《懷孕天書》、《100 保健湯水 ● 茶飲》、《滋補甜品店》、《女性調補天書》、《強身 ● 治病孩子飲食全書》、《上班族每天養生法～飲出好體質》、《養護好體質～手術前後飲食自療》。著作迅即再版，成績斐然。

序

有研究指香港每 6 對夫婦就有一對不育，不孕不育病發率在現代醫學水平不斷提高情況下，不但沒有降低，反而明顯增加，情況令人關注。引致不育的原因非常多，除了男女雙方身體出了問題，不良的生活習慣及起居飲食都可能增加了不育的發病率；同時大都市生活節奏太快，工作壓力過大，精神長期處於緊張狀態，也可能間接引致男女性欲低下、性功能障礙及生殖力下降。

當一對夫婦久久未能成孕，最重要是去作一些常規的體檢，以及功能型的檢查，例如甲狀腺功能、生育功能、卵巢功能等各方面的常規檢查。如果兩夫婦都未作過任何體檢，卻試圖只用民間偏方、中醫藥膳或食療來治療不育症，這是起不到太大作用的。

中、西醫學對治療不育症各有所長，亦各具不同療效。倘若夫婦兩人都排除了卵巢、子宮病變、輸卵管、睪丸病變、精子稀少等各種不利懷孕的因素，檢驗亦找不出不育成因，也許透過改善不良的生活習慣，包括戒煙、節制飲酒、作息定時和恆常運動，加上充足營養和均衡的飲食調理，有助實現成孕。

本書採用的，全是天然食物製成的食療，少用峻烈中藥，全書共有 50 多款美味可口的小菜、粉麵、粥品、茶飲、湯品等，用食補的方法進行調理，對女性可以暖宮補血、補脾腎；對男性可補腎強精、增加精子活力，提升夫妻間的性功能。全書材料簡單易於購買，美味可口，並有詳細介紹每種食材所起的作用，以及每款食療的功效、宜忌等，希望讀者透過神奇的食材力量，找到適合自己應用的食療，調理好體質。

當然讀者可能依足本書的食譜進行調理，卻未必一定能夠成功懷孕，皆因不孕很多時會受到年齡、情緒、居住環境，甚或一些被忽略的生活細節等所影響；不過只要朝着調養好身體這個方向走，實現成孕也並非難以達到之事。

目錄

增強男女性能力飲品

暖宮補血、補腎強精之家常小菜

促進生育之主食

增強卵子、精子活力飲品甜點

contents

Daily dishes for boosting fertility

Staples for both sexes

Drinks and desserts for healthy sperms and eggs

認識不育

有研究表示，夫妻性生活正常，又沒有避孕，倘若 12 個月內仍未能受孕，醫學上界定為不育。以往生兒育女是女人的天職，未能受孕會被認為問題主因在於女方；現代醫學研究卻顯示，約有三成不育原因在於男方，而女方也只佔三成，其餘則屬男女雙方引致或原因不明。

大都市生活壓力大，精神緊張，焦慮症、抑鬱症患者比比皆是，情緒病令夫妻間的性生活出現障礙，性欲下降，生殖力也跟着下降；此外，電子輻射污染、空氣、食水及食物污染直接影響受孕。市面大量用激素飼養的家畜、家禽也越來越多，食物污染引致性早熟，令未婚先孕或意外懷孕機會大增，更使人工流產率日益上升；過早有性生活令致通過性接觸而感染的疾病日漸增加，也讓不孕不育症更為普遍；這些都可能是男女雙方不孕不育而又難宣之於口的主要原因之一。

所以計劃生育寶寶的夫妻，必須先調整好自己情緒，身體有毛病者要及早治理好，切勿諱疾忌醫。除了要戒煙、戒酒，應注意作息有定時，培養良好的飲食習慣亦非常重要，因為充足有益的營養可以幫助實現成孕。

紅菜頭汁

　　日常飲食要注意補充各種維他命、礦物質及微量元素，各種營養成分要合理地吸收；因為任何成分缺少了都會影響精子的質量和活力，也影響女性荷爾蒙的分泌。但這些營養成分也並非越多越好，例如不少男士以為吃了鹿尾羓或牛鞭會吸收更多雄激素，因而能夠增加性能力；女士們以為吃名貴燕窩、雪蛤膏、紫河車等補品可增加更多雌激素以延緩衰老，保持容顏青春。其實這類食物如果食用過量，可能會擾亂了身體荷爾蒙分泌，使激素水平失衡，導致不孕，甚至患上前列腺癌、乳癌、卵巢癌等。所以最好多從天然植物及素食中吸收，如喝黑豆漿、紅石榴汁、紅菜頭汁等，多吃豆類及各種豆製品，以及各種新鮮蔬果、菌類、堅果，從食物中攝取的「植物雌激素」更能讓人放心。只要食得健康，養護好體質，成孕的機會亦會大增。

男性不育及改善方法

男性不育的最主要成因是健康精子的數量不足、不夠活躍。當中不少人更會因工作、生活上各種繁瑣事務令自己情緒不安、煩躁易怒、焦慮抑鬱、精神不振，這些都容易造成陽痿不舉的問題，故放鬆心情至為重要。日常生活節奏宜放慢，不要因不育給自己添加無形壓力，造成肝氣鬱結。飲食作息更需要調節，只要有恆心去改善體質，就有望增加成孕機會。

不育與年齡有莫大關係，有研究指 45 歲以上男性的精子數量和活動能力下降，甚至可能有變畸形的精子出現。有些男人過了 40 歲以後，體能方面明顯走下坡，精神方面亦會出現注意力不集中、健忘等症狀，性生活方面也會表現性欲降低情況；究其原因，雄激素缺乏是禍首，如能合理地進食能刺激雄激素分泌的食物，大可改善這方面的情況。

有助刺激雄激素分泌的食物

1.含鋅量高的食物：如肝臟、蠔肉、牛肉、雞肉、牛奶、蛋黃、海鮮、貝類、海帶、豆類、花生、小米、小麥胚芽、蘿蔔、薯仔、南瓜子、核桃、松子、芝麻等。

2.含精氨酸的食物：如鱔魚、大比目魚、龍蝦、三文魚、蝦、螺、金槍魚、泥鰍、海參、墨魚、鱘魚、雞肉、豆腐、紫菜、豌豆等。

3.含鈣的食物：如小魚乾、蝦皮、鹹蛋、蛋黃、乳製品、大豆、海帶、芝麻醬等。

4.富含維他命的食物：維他命 A、維他命 E 和維他命 C 都有助於延緩衰老和避免性功能衰退，它們大多存在於新鮮蔬菜、水果中。

5.最能增進男性活力的食物：韭菜、蝦肉、海參、蠔肉、蛤蜊、蜆肉、南瓜子、核桃肉、杞子、芡實、番茄等。

男性不育及改善方法

温馨提示

　　放射性物質、高溫及毒物等會影響男性的性功能，及影響精子質量和活力，例如穿着剛從洗衣店用乾洗劑清洗取回來的衣物、泡熱水浴、長期接觸放射性物質，在高溫爐火旁工作等，都會使睪丸溫度升高影響生育。煙酒過多及過食高脂肪肥膩食物，也會影響男性的性功能。

生活小知識

　　有多種植物已被科學家研究出能有效淨化及過濾髒空氣，這些植物可吸收空氣中有害煤煙、放射性物質、化學物質、有毒粉塵等，屬天然空氣清新劑，如仙人掌、虎尾蘭、蘆薈、黃金葛、吊蘭、富貴竹、白掌、巴西鐵樹、袖珍椰子等。

　　日常在家居或工作間多種植這類能淨化空氣的植物，可增加放氧量，減少吸入污濁空氣及輻射、甲醛、臭氧等物質，身體毒素減少，對增強精子數量及活力有一定幫助。

女性不孕及改善方法

　　現代男女趨向晚婚，有研究指女性到了 30 歲，一半卵子屬不正常，35 歲時七成，40 歲則有九成卵子有問題；所以，年紀愈大，受孕機會愈低。即使醫學科技再先進，但年齡始終是不能成孕的主因，高齡婦女在人工受孕方面成功例子往往比年輕女性為低。

　　很多中醫師認為，子宮就像是胎兒的暖房，是孕育新生命的搖籃。如果子宮內冰冷，那麼胎兒就無法生長，因此有「宮寒不孕」的說法；所以女性在月經來前幾天至來經期間，要避免食用生冷寒涼食物及冰凍冷飲，宜用溫補食物，暖飲暖食。每次月經結束後，不妨檢討當月的經血狀況，像經血顏色的深淺、經量多或少、是否有經痛、以及來經有沒有規則性等，從而觀察當月子宮、卵巢是否正常，或是還欠缺了甚麼？這是想懷孕的女性必須重視的月經訊息。

　　卵子與精子需要的營養非常均勻，如果過於偏食，或曾經長期減肥，體內缺乏某些營養，卵子就不會優良，也就不容易受孕。例如蛋白質不夠會造成經血量少，影響受孕機會；富含鐵質的食物亦需要適量補充。

有助孕功效的食物

1.含豐富蛋白質及鈣質的食物：如牛奶、小魚乾、蝦米、豆漿、豆乾、雞蛋、乳酪、無花果、黑芝麻及各種肉類等。

2.含鐵量高的食物：如黑木耳、金針、豬肝、牛肉、羊肉、蝦米、黑芝麻、花生、榛子、黑豆、紅豆、腐竹、腐皮、燕麥、粟米片、圓肉、金針、葡萄乾、棗類、菠菜、西蘭花、紅菜頭等。

3.含豐富膠質的食物：如海參、花膠、雪耳、黃耳、黑木耳、海藻、豬腳筋、雞腳、豬皮等。

4.含豐富維他命 E、B$_6$ 的食物：如核桃、黑芝麻、大豆、紅豆、黑豆、三角豆、花生、腰果、糙米等。

5.含鋅量高的食物：如肝臟、蠔肉、牛肉、雞肉、牛奶、蛋黃、海鮮、貝類、海帶、豆類、花生、小米、小麥胚芽、蘿蔔、薯仔、南瓜子、核桃、松子、芝麻等。

6.含豐富維他命 C、抗氧化的生果：如檸檬、蘋果、百香果、木瓜、葡萄、藍莓、紅石榴等。

温馨提示

　　長期飲咖啡也會影響受孕機會；有研究指每天喝一杯以上咖啡的女性，懷孕的可能性只是不喝此種飲料者的一半。因此，女性如果打算懷孕，就應該少飲些咖啡。

　　此外，婦女如白帶過多易受細菌感染而產生炎症，如缺乏合適治療，易造成輸卵管及盆腔出現炎症，使管道堵塞精子難以通過，更可能出現免疫排斥現象，大大影響受孕機會。故一旦陰道分泌物不正常，必須耐心把病徹底治好。

生活小知識

　　番茄、紅蘿蔔、蘋果、南瓜、香蕉、黑木耳、西蘭花、蘆筍、大豆芽菜、捲心菜（椰菜）、海帶、海藻、綠茶等含有果膠、分解酶、單寧酸等物質，有抗輻射損傷的功能，並能溶解沉澱於細胞內的毒素，使之隨尿液、糞便中排走。身體毒素減少，對成孕大有裨益。

　　此外，海參、花膠能補腎益精，滋補養顏，這兩種食物既能滋補腎陰，又能補益腎陽，屬陰陽雙補之品，男女皆適合食用。

　　花膠及海參含有多種微量元素、膠原蛋白及黏多糖等物質，是人體補充、合成蛋白質的主要原料，人體易於吸收利用，可有效提高身體免疫力，抑制癌細胞，對人體多種生理活動，尤其是生殖功能作用突出，大大提升性功能。對一些遲婚，有意孕育寶寶之男女，日常不妨多食用這類有益健康的食物，及早實現成孕。

花膠響螺鵪鶉湯

女：助孕、暖宮湯水

暖宮對女性的重要性

　　女性宮寒除了容易手腳冰冷外，更容易引起月經失調、痛經、臉上長黃褐斑、性欲低下等不良反應，甚至可能會因此造成不孕，或是胎兒出現發育遲緩等問題；因此，暖宮對女性來說非常重要！

　　其實子宮與卵巢都是身體內臟之一，需要足夠營養才能運作暢順，雖然子宮相對心臟、脾胃等器官對人體來說重要性較低，但一旦營養不足，首當其衝就是生殖系統；因此均衡飲食是養宮、暖宮的基本條件。

　　要暖子宮就要避免子宮被寒濕之氣凝聚，導致宮寒。女性要注意保暖，來經期間應盡量避免食用生冷食物及冰品冷飲，日常亦不宜過食生冷寒涼的食物。如要暖宮，可以多用一些溫性及對子宮有益的生薑、胡椒、洋葱、羅勒葉、豬肝、牛肉、羊肉、雞肉、海參、蠔豉等食物以溫暖子宮，進而增加受孕機會。

胡椒黑豆豬肚湯

Pork tripe soup with white peppercorns and black beans

女：助孕、暖宮湯水

胡椒

黑豆

認識主料

胡椒：胡椒分白胡椒和黑胡椒，白胡椒是完全成熟的果實，而黑胡椒是未成熟的果實。黑胡椒的辣味比白胡椒強烈，香中帶辣，祛腥提味；白胡椒則能健胃、散寒、增進食欲、助消化，促發汗，還可以改善女性白帶異常及癲癇症。

黑豆：黑豆含豐富的優質蛋白、維他命 B 群及維他命 E，還有鐵、鈣、磷、鉀、鎂及花青素等，這些都是構成人體細胞基本物質，有保護子宮正常發育的功效，女性宜多食黑豆或飲黑豆漿。

材料（3~4 人量）　　　調味料

白胡椒粒 5 克　　　　　海鹽半茶匙

青仁黑豆 30 克

豬肚 1 個

瘦肉 200 克

芫茜碎 1 棵份量

做法

1. 青仁黑豆浸洗；豬肚用鹽及生粉反覆搓洗淨，出水後切塊；瘦肉切厚塊，出水。

2. 全部材料和 1,500 毫升水先用大火煮滾，再用中小火煮 2 小時，調味及灑入芫茜碎即成。

食療功效

溫中散寒、健胃補虛。

飲食宜忌

本品香味濃郁、增加食欲。適合虛勞瘦弱、食欲不振、氣血虧虛、白帶過多、泄瀉下痢者。痰濕內盛、陰虛火旺及外感發熱者不宜。

● 建議每月食用 1~2 次

女：助孕、暖宮湯水

木瓜花生排骨湯

Pork rib soup with papaya and peanuts

女：助孕、暖宮湯水

材料（3~4 人量）
青木瓜 1 個
花生 30 克
陳皮 1 個
紅棗 4 粒
排骨 300 克

調味料
海鹽半茶匙

做法

1. 青木瓜去皮洗淨，切塊；花生、陳皮浸洗；紅棗去核；排骨出水。
2. 全部材料和 1,500 毫升水用大火煮滾，轉用中小火煮 1 小時，調味即成。

青木瓜：青木瓜一般都帶有點苦澀味，果漿味也比較濃。有助消化、潤滑肌膚、分解體內脂肪、刺激女性荷爾蒙分泌、刺激乳腺激素等功效。並具有防治高血壓、腎炎、便秘和助消化、治胃病，對人體有促進新陳代謝和抗衰老的作用，還有美容護膚養顏的功效。

花生：花生有補益脾胃、潤肺健腦等功效，所含維他命 E、K 及一定量的鋅，有助增強記憶、延緩衰老、滋潤肌膚，更有止血、催奶及增加精子數量及精液生成等作用。

健脾補血、促進消化。

本品鮮味可口，老少咸宜。適合脾胃虛弱、病後體虛、營養不良、食欲不振及荷爾蒙分泌不足者。一般人群可食，但孕婦不宜。

● **建議每星期服用 1~2 次**

紅菜頭腰果肉片湯

Lean pork soup with beetroot and cashews

女：助孕、暖宮湯水

材料（2~3 人量）
紅菜頭 1 個
番茄 2 個
腰果 30 克
瘦肉 250 克

調味料
海鹽 1/4 茶匙

做法

1. 紅菜頭、番茄去皮，切塊；腰果沖淨；瘦肉切厚片，出水。

2. 全部材料和 1 公升水用中小火煮 1 小時，調味即成。

認識主料

紅菜頭：紅菜頭能護肝、補血、降血壓、保護心臟、抗癌。紅菜頭含維他命 B_{12} 及鐵質，是婦女補血的最佳天然營養品。菜汁中含有微量硼元素，有助提高男性睪丸甾酮分泌量，強化肌肉，硼還有改善腦功能、提高反應能力的作用。

腰果：腰果能補充體力、消除疲勞，提供能量，適合易疲倦的人食用。並有潤膚美容、延緩衰老、增進性欲等功效。

食療功效

保護血管、增進性欲。

飲食宜忌

本品香甜美味、顏色吸引。適合血虛、面色萎黃、精神不振、性欲淡漠者。低血壓者忌服。

● 建議一星期服用 1~2 次

羅勒葉牛肉湯

女：助孕、暖宮湯水

材料（2 人量）
羅勒葉 50 克
生薑 3 片
牛肉 250 克

調味料
海鹽 1/4 茶匙

做法

1. 羅勒葉洗淨，去梗；牛肉洗淨，切薄片。
2. 燒熱 600 毫升水，加入牛肉、生薑，滾 10 分鐘，再加入羅勒葉，滾 3 分鐘，加調味即成。

認識主料

羅勒葉：羅勒葉又名九層塔。羅勒葉對於調理體質、改善血液循環、增強免疫系統有很好的功效。羅勒葉可增強胃屏障、抗胃潰瘍，亦能促進卵泡成熟排出。

牛肉：牛肉有補中益氣、滋養脾胃、強健筋骨、化痰息風等功效。牛肉蛋白質含量高，而脂肪含量低，並可提供更多的鐵和鋅，可提高免疫系統的功能，還富含葉酸，可預防新生兒缺陷，並可幫助保持男性的精子質量。

食療功效

滋養脾胃、促進卵泡成熟。

飲食宜忌

本品香濃美味，簡單易煮。適合
貧血、頭痛、月經不調、精神沮
喪、氣短體虛及不易受孕者。氣
虛血燥及有皮膚舊患者不宜。

● 建議每星期服用 1~2 次

蟲草花杞子乳鴿湯

Squab soup with cordyceps flowers and goji berries

材料（2~3 人量）

蟲草花 6 克

杞子 5 克

生薑 3 片

紅棗 6 粒

乳鴿 1 隻

調味料

海鹽 1/4 茶匙

做法

1. 蟲草花、杞子浸洗；紅棗去核；
乳鴿劏洗淨，斬件後出水。

2. 全部材料和 1 公升水用中小火
煮 1 小時，調味即成。

滋補肝腎、強壯體質。

飲食宜忌

本品甘香味美，滋補強身。適
合腎虛體弱、體力透支、精氣
不足、心神不寧、陽痿遺精及
性冷感者。外感發熱者不宜。

● **建議每隔 3 天服食 1 次**

認識主料

蟲草花：蟲草花有補肺腎、止咳嗽、
益虛損、扶精氣之功，適用於肺腎
兩虛，精氣不足，陽痿遺精，咳嗽
氣短，自汗盜汗，腰膝酸軟等。

乳鴿：白鴿的繁殖力很強，性欲極
強，雌雄交配很頻密，這是由於白
鴿的性激素分泌特別旺盛所致；所
以人們把白鴿作為扶助陽氣強身補
品，認為它具有補益腎氣、強壯性
機能的作用。

花生杞子竹絲雞湯

Silkie chicken soup with peanuts and goji berries

女：助孕、暖宮湯水

材料（2~3 人量）
花生 30 克
杞子 6 克
紅棗 6 粒
生薑 3 片
竹絲雞半隻

調味料
海鹽 1/4 茶匙

做法
1. 花生、杞子浸洗；紅棗去核；竹絲雞劏洗淨，斬大件後出水。
2. 全部材料和 1 公升水用中小火煮 1 小時，調味即成。

認識主料

竹絲雞：有滋陰清熱、補肝益腎、健脾止瀉等功效。並可提高生理機能、延緩衰老、強筋健骨、對防治骨質疏鬆、佝僂病、婦女缺鐵性貧血症等有明顯功效。

食療功效

健脾和胃、補益肝腎。

飲食宜忌

本品清甜美味，補而不燥。適合體虛
血虧、記憶力減退、脾胃不健、月經
不調者。外感發熱及肝陽上亢者不宜。

● 建議每星期服用 1~2 次

淮杞海參燉土雞

Double-steamed chicken soup with Huai Shan and sea cucumber

女：助孕、暖宮湯水

食療功效

補腎益精、養血潤燥。

飲食宜忌

本品清香味美、滋補強壯。適合肝腎虧虛、脾虛泄瀉、營養不良、月經不調、遺精、陽痿者。濕熱體質及外感未清者不宜。

● 建議每隔 3 天服食 1 次。

材料（2~3 人量）

淮山 30 克
杞子 5 克
浸發海參 200 克
紅棗 4 粒
生薑 3 片
土雞半隻

調味料
海鹽 1/4 茶匙

做法

1. 淮山、杞子浸洗；浸發海參切塊，與斬件土雞同出水；紅棗去核。

2. 全部材料放入燉盅內，注入 600 毫升開水，隔水燉 2 小時，調味即成。

認識主料

淮山：淮山能健脾、厚腸胃、補肺益腎。淮山富含植物雌激素，可刺激卵巢排卵，有助改善黃體期過短的問題。對脾虛泄瀉、久痢、遺精、帶下、子宮脫垂等均有療效。

土雞：土雞指放養在山野林間、果園的肉雞。雞肉含豐富的蛋白質及不飽和脂肪酸。對心血管病患者及體質虛弱、病後或產後人士最適合。對虛勞瘦弱、骨蒸潮熱、脾虛泄瀉、婦女崩漏、白帶及男子遺精等都有益。

海參：海參能補腎益精、養血潤燥、止血消炎。除了蛋白質豐富，還含有多種海參素、黏多糖、礦物質等，有助促進生殖功能，且完全不含膽固醇。海參自古被視為補腎益精及壯陽療痿之珍品，對體弱遺精、脾虛血弱及神經衰弱人士頗為有益。

鮮蠔蘿蔔絲湯

Oyster and radish soup

材料（2~3 人量）

急凍中等鮮蠔 150 克
白蘿蔔 1 條
薑絲 1 湯匙
羅勒葉 2 株份量
高湯 500 毫升

調味料

海鹽 1/4 茶匙
胡椒粉適量

做法

1. 蠔肉解凍，出水撈起；白蘿蔔去皮，切絲。
2. 用少許油爆香薑絲，倒入高湯及蘿蔔絲，大火煮 15 分鐘，加入蠔肉、羅勒葉煮 5 分鐘，調味即可供食。

食療功效

消食化痰、調中補虛。

飲食宜忌

本品清甜美味，滋補強壯。適合胃熱、便秘、流虛汗、精液不足者。痛風患者不宜。

● 建議每星期服 1~2 次。

白蘿蔔：蘿蔔能清涼止渴、利尿、化痰、助消化、清肺胃熱毒。蘿蔔含有大量維他命和礦物質，能增強人體免疫力，所含的纖維素可加強胃腸蠕動，促進排便。

鮮蠔：蠔肉有調中補虛、滋陰養血的功效。含人體所需十幾種氨基酸、礦物質、多種維他命等，含鋅量特別高，而精液之生成必須要有足夠的鋅才可以，故蠔肉有壯陽的功效。

● 女：助孕、暖宮湯水 ●

姫松茸杏仁鵪鶉湯

Quail soup with Himematsutake
and almonds

材料（2~3 人量）

姬松茸 10 克

杏仁 10 克

生薑 3 片

紅棗 6 粒

鷓鴣 1 隻

調味料

海鹽半茶匙

做法

1. 姬松茸浸軟；杏仁沖淨；紅棗去核；鷓鴣劏洗淨，斬件後出水。

2. 全部材料和 1,500 毫升水用大火煮滾，再轉用中小火煮 1.5 小時，調味即成。

食療功效

開胃化痰、扶正補虛。

飲食宜忌

本品清香味美，滋補有益。適合氣血不足、痰多咳喘、體弱多病、神疲乏力者。外感未清及對菇類敏感者不宜。

● 建議每星期服 1~2 次。

認識主料

姬松茸：姬松茸有健腦益腎、扶正補虛之效，是體弱多病者的補身佳品。姬松茸能降膽固醇和血脂、降血糖，能提升血管健康、保護肝腎、增強體能、延緩衰老、抗過敏反應，更能抑制癌細胞生長。

鷓鴣：鷓鴣是一種優質的滋補營養品。鷓鴣骨細肉厚，肉嫩味鮮，營養豐富，具有高蛋白、低脂肪、低膽固醇的營養特性。功能滋養補虛、開胃化痰。且有斂汗作用，補而不燥，特別適合虛火盛、虛不受補人士。

百合蛋花湯

Egg drop soup with lily bulbs

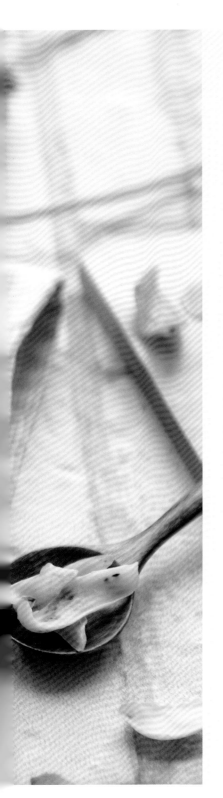

材料（2 人量）　　調味料

百合 30 克　　　海鹽 1/4 茶匙

火腿 1 塊

薑絲 2 茶匙

雞蛋 1 個

高湯 600 毫升

做法

1. 百合浸洗；火腿切絲；雞蛋打散。
2. 燒熱高湯，加入百合、薑絲煮 20 分鐘，最後加入蛋漿、火腿絲及調味即成。

食療功效

補中益氣、寧心安神。

飲食宜忌

本品清香可口，簡單易煮。適合虛煩不安、失眠多夢、食欲欠佳及腦力減退者。一般人士可服。

● **建議每隔 3 天食用 1 次。**

認 識 主 料

雞蛋：雞蛋幾乎含有人體必需的所有營養物質，富含 DHA 和卵磷脂、卵黃素，對神經系統和身體發育有利，能健腦益智，改善記憶力，並促進肝細胞再生。

男：增強精子活力、補腎湯水

增強精子活力要補腎

經常吸煙、喝酒、長時間久坐及操作電腦、工作繁忙、精神緊張、生活及飲食無規律、性生活頻密的男性，最易出現腎虛症狀。腎功能失常會導致性功能異常，男性最常見的是出現性欲下降、陽萎、遺精、滑精、早泄、精子活動力減弱、不育等。

腎虛會影響生育；故如想孕育寶寶，首要戒煙、少飲酒，更要改善作息時間，治理好個別器官的疾病以防累及腎臟。如發現經常莫名疲勞、頻打哈欠、腰痠背痛、夜尿頻、記憶力下降等情況，就要及早作出調理。日常多飲用一些補腎的湯水，有助精子的數量和活力得以改善，從而增加受孕的機會。

南瓜奶油濃湯

Cream of pumpkin soup

男：增強精子活力、補腎湯水

南瓜

洋葱

認識主料

南瓜：南瓜能補中益氣、消炎止痛、解毒去蟲，更是極好的 Omega-3 脂肪酸來源，南瓜籽和果瓤都包含大量 Omega-3。含有 Omega 3 脂肪酸豐富的食物，可以增加精子數量及生殖系統的血流量。

洋葱：洋葱有祛痰、利尿、健胃潤腸、解毒殺菌等功能，並可提高胃腸道張力、增加消化道分泌作用。洋葱含鈣量高，能防止骨質疏鬆，經常食用對患高血壓、高血脂和心腦血管人士都有幫助。

材料（2~3 人量）

南瓜 200 克

洋葱半個

高湯 500 毫升

鮮忌廉 200 毫升

牛油 25 克

麵粉 15 克

蒜頭 2 瓣

歐芹碎 1 湯匙（或免）

調味料

海鹽半茶匙

做法

1. 南瓜去皮，洗淨切粒；洋葱去衣切粒；蒜頭拍碎。

2. 牛油加熱，放入洋葱、蒜蓉炒香，加入南瓜炒至變軟，水分溢出，灑下麵粉使材料收水，然後注入高湯，拌勻，關火。

3. 將材料放入攪拌機中打成蓉，用小火煮滾，關火，加入鮮忌廉攪拌及調味，灑入歐芹碎即成。

食療功效

補中益氣、健胃潤腸。

飲食宜忌

本品香滑美味，老少可飲。適合消化力弱、食欲不振、神疲乏力、精子數量不足者。濕重人士宜少食。

● 建議每月食用 1~2 次

核桃子雞湯

Walnut cockerel soup

男：增強精子活力、補腎湯水

材料（2~3 人量）

核桃肉 30 克

南棗 4 粒

生薑 3 片

少爺雞 1 隻

調味料

海鹽半茶匙

做法

1. 核桃肉、南棗沖淨；少爺雞劏洗淨，斬大件後出水。

2. 將全部材料放入煲內，注入 1,500 毫升水，用大火煮滾後，改用中小火煮 1.5 小時，調味即成。

認識主料

核桃：核桃能補腎固精、温肺定喘。其中的 Omega-3 脂肪酸和維他命 E 可以促進子宮內膜健康，甚至還能降低罹患前列腺癌和乳癌的風險。每天吃一些核桃，可以提升精子的活力。

南棗：南棗是大棗的加工品，看上去很像黑棗，但黑棗較胖而圓，南棗較瘦而長，南棗沒有黑棗的膩滯，補而不燥，具有滋腎養肝、健腦安神、養顏烏髮等功效，同時對腹瀉、小兒遺尿等症甚有裨益。

好的南棗應隻形大，肉質堅實，甜而軟糯，乾身，皮紋細緻清晰，搖它可聽到棗核在內被搖動作響的小聲音。

健脾補血、促進消化。

本品鮮味可口、滋補強身。適合體質虛弱、容易咳喘、頭暈心悸、小便頻數、遺精、精少精冷者。外感發熱者及有皮膚舊患者不宜用少爺雞。

● 建議每星期服用 1~2 次

白果腐竹冬菇蠔豉湯

Dried oyster soup with gingkoes,
beancurd skin and shiitake mushrooms

材料（2~3 人量）

白果 15 粒
鮮腐竹半張
冬菇 4 朵
蠔豉 50 克
陳皮 1 塊

調味料
海鹽半茶匙

做法

1. 白果去芯，陳皮浸軟，冬菇浸軟後去蒂，蠔豉浸洗乾淨。

2. 全部材料和 1,500 毫升水用大火煮滾，轉用中小火煮 1.5 小時，調味即可供食。

食療功效

健脾益胃、滋補壯陽。

飲食宜忌

本品濃香美味，滋補而不燥火。適合肺弱咳喘、骨質疏鬆、遺精、白帶、精力衰退者。脾胃虛寒者少食。
● **建議每星期服用 1~2 次**

認識主料

白果：白果即銀杏，能潤肺、定喘、澀精、止帶。適合患有哮喘、痰嗽、白帶、白濁、遺精、淋病、小便頻數者食用。白果還具有通暢血管、改善大腦功能、延緩老年人大腦衰老、增強記憶能力和改善腦供血不足等功效。

蠔豉：蠔豉具有滋陰養血、補益五臟、壯陽及美容功效。蠔豉含鋅量很高，精液的生成必須有足夠的鋅才可以。此外，發育時期的青少年及婦女必須吸收足夠的鋅元素，身體發育才好，卵巢才能產生黃體素，幫助排卵。

動、增強清季活力、補賢湯水。

紫菜番茄蛋花湯

Egg drop soup with laver and tomatoes

食療功效

健胃消食、清熱利尿。

飲食宜忌

本品清香可口，簡單易煮。適合貧血、消化不良、小便不暢、肥胖水腫及不育不孕者。一般人士可服，甲亢患者不宜用紫菜。

● 建議每星期服用 1~2 次

材料（2~3 人量）

紫菜 5 克

番茄 2 個

雞蛋 2 個

葱絲 1 湯匙

薑絲 2 茶匙

調味料

海鹽半茶匙

做法

1. 紫菜浸洗；番茄去皮，切塊；雞蛋打散。

2. 燒熱 600 毫升水，加入紫菜、番茄、薑絲，滾 15 分鐘，加入調味料及打散蛋漿，滾起，灑入葱絲即成。

認識主料

紫菜：紫菜能化痰軟堅，清熱利尿。紫菜含碘量很高，可用於治療因缺碘引起的「甲狀腺腫大」，並有軟堅散結功能，對其他鬱結積塊也有作用；紫菜富含膽鹼和鈣、鐵、能增強記憶、治療貧血。

番茄：番茄所含的茄紅素能預防細胞受損，並協助人體對抗癌症及老化現象，提升免疫力。茄紅素有保護生殖器官的功能，因茄紅素進入人體後，會大量分佈在攝護腺、睪丸及上腎上腺，形成保護作用。

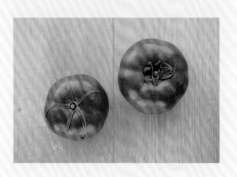

大豆芽菜淡菜湯

Dried mussel soup with soybean sprouts

男：增強精子活力、補腎湯水

大豆芽菜

淡菜

認識主料

大豆芽菜：與黃豆一樣，大豆芽也有滋潤清熱、利尿解毒之效，同時含有植物性荷爾蒙，對男女均有益。黃豆在發芽過程中，由於酶的作用，會產生更多礦物質元素被釋放出來，故會增加礦物質在人體中的利用率。大豆芽菜更有消除皮膚疣的作用。

淡菜：淡菜俗稱旺菜，具有補虛益精、溫腎散寒的功效，也是滋陰平肝的營養食品。適用於男子性功能障礙、遺精、陽痿、房勞、消渴等症。男子常食可強壯身體增強性功能、提高精子的質量。對防治腎功能虛損、男性陽萎、早洩、精力不足、婦女陰虛白帶很有效。

男：增強精子活力、補腎湯水

材料（2~3 人量）

大豆芽菜 200 克

冬菇仔 15 克

淡菜 50 克

甘筍 1 條

生薑 3 片

調味料

海鹽半茶匙

做法

1. 大豆芽菜去尾，白鑊炒乾；冬菇仔浸
   洗，去蒂；淡菜浸洗後出水；甘筍去
   皮後切塊。

2. 燒熱 1,500 毫升水，加入全部材料，
   用中大火煮 1 小時，調味即可供食。

食療功效

滋陰補腎、提高性能力。

飲食宜忌

本品清香味美、滋潤養顏。適合皮膚長
斑長疣、腎虛水腫、高血壓、高血脂、
遺精、白帶及精力不足者。一般人士可
服，痛風患者不宜。

● 建議每星期服 1~2 次。

男：增強精子活力、補腎湯水

大蒜蜆肉湯

Garlic clam soup

材料（2人量）

蒜頭 6~8 瓣

韭黃 5~6 棵

新鮮大蜆 250 克

紅椒碎 1 茶匙

高湯 500 毫升

調味料

海鹽、胡椒粉、米酒適量

做法

1. 蒜頭去衣，略拍；韭黃洗淨，切段；大蜆用清水浸半天吐砂。

2. 煮滾高湯及蒜頭，放入大蜆，待蜆殼張開，灒酒加調味料，灑入韭黃段及紅椒碎，滾起即成。

食療功效

清熱解毒、保護肝臟。

飲食宜忌

本品鮮味可口、滋補有益。適合食慾不振、體倦乏力、腳氣水腫、精液不足、糖尿病、血脂高、血壓高者。痛風者不宜食貝類食物。

● 建議每星期服食 1~2 次

認識主料

蒜頭：蒜頭含有的肌酸酐，是參與肌肉活動不可缺少的成分，對精液的生成也有作用，可使精子數量大增。它富含硒，可提高精蟲的數量和活動力。蒜頭還能促進新陳代謝，降低膽固醇和甘油三酯的含量，並有降血壓、降血糖的作用，同時可改善因腎氣不足而引發的渾身無力症狀。

蜆肉：蜆肉能清熱、利濕、解毒。含有多種胺基酸，可提供肝細胞使用，蜆肉含有豐富的鈣質和肝醣，可以讓身體能量增加，促進造血功能及肝功能的恢復，對肝臟健康很有益。

鮮蝦杞子童子雞湯

Cockerel soup with shrimps and goji berries

男：增強精子活力、補腎湯水

材料（2 人量）

鮮蝦 100 克

杞子 5 克

生薑 3 片

紅棗 6 粒

少爺雞半隻

調味料

海鹽 1/4 茶匙

做法

1. 鮮蝦去殼，挑去腸臟；杞子浸洗；紅棗去核；少爺雞劏洗淨，斬件後出水。

2. 雞件、紅棗、生薑和 1,500 毫升水用大火煮滾，改用中小火煮 1.5 小時，加入杞子、鮮蝦滾 5 分鐘，調味即成。

認識主料

鮮蝦：蝦肉能補氣健胃、暖腎壯陽。對脾胃虛弱、神經衰弱、腎虛陽痿、腰膝痠軟、神疲乏力者有益。蝦含鋅量高，能增加精子活力，及提升性功能。

少爺雞：少爺雞即童子雞。少爺雞肉蛋白質的含量比例較高，種類多，而且消化率高，容易被人體吸收利用，有增強體力、強壯身體的作用。對虛勞瘦弱、脾虛泄瀉、月經不調、小便頻數、遺精、精少精冷等有益。

温腎壯陽、益氣助孕。

本品鮮美可口，滋補強身。適合虛勞瘦弱、腎虛早洩、陽痿、崩漏帶下、小便頻繁者。痛風患者及外感未清者不宜。

● 建議每星期服用 1~2 次

紫淮山粟米濃湯

Cream of purple yam soup with sweet corn

男：增強精子活力、補腎湯水

材料（2 人量）
鮮紫淮山 100 克
急凍粟米粒 20 克
牛奶 500 毫升
蛋白 2 個
生粉水 2 湯匙

調味料
海鹽半茶匙
胡椒粉少許

做法
1. 紫淮山去皮，洗淨切塊，與 1 杯 (250 毫升) 鮮牛奶放攪拌器打成蓉。
2. 將紫淮山蓉、粟米粒及其餘牛奶用慢火加熱，滾起後加生粉水及調味料，邊煮邊攪，最後加入蛋白，滾起可供食。（紫淮山含花青素，故煮好後部分蛋白可能會轉淺藍色。）

男：增強精子活力、補腎湯水

食療功效

健脾補虛、強壯體質。

飲食宜忌

本品香滑美味、滋陰補虛。適合工作壓力大、
久病體虛、氣血不足、營養不良、不孕不育者。
一般人士可服。

● 建議每星期服用 1~2 次

認識主料

紫淮山：紫淮山含大量花青素，有抗
氧化、美容養顏的功效，其所含的薯
蕷皂素能促進內分泌荷爾蒙合成，對
卵子生長、乳腺發育等很有幫助。對
防治高血壓、肝炎、癌症、腦力衰退
等有良好效果，常食更有助預防女性
子宮下垂，男性遺精。（如買不到紫
淮山，可以用紫淮山粉 30 克代替。）

牛奶：牛奶具有補虛損、益肺胃、生
津潤腸之功效。牛奶同時能刺激紅血
球活化，提升免疫力，促使卵子更有
活力，增加成孕的機率。

杞子

鮮石斛

鮑魚

鮮石斛杞子鮑魚湯

Abalone soup with fresh Shi Hu and goji berries

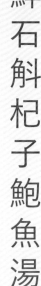

男：增強精子活力、補腎湯水

認識主料

杞子：杞子能滋補肝腎、益精明目。適用於虛勞精虧，腰膝痠痛、眩暈耳鳴、內熱消渴、血虛痿黃、目昏不明及遺精等症。杞子還有延緩衰老、抗脂肪肝、調節血脂、血糖、促進造血功能等作用。

鮮石斛：鐵皮鮮石斛能滋養津液、養胃陰，潤膚養顏，並能保護女性卵巢，並具有滋潤臟腑及腦髓骨骼的作用，有助維持正常的生長發育與生殖功能活動。

鮑魚：鮑魚含有豐富的蛋白質，還有較多的鈣、鐵、碘和維他命 A 等營養元素。鮑魚具滋陰作用，補而不燥，對夜尿頻、氣虛哮喘、血壓不穩、精神難以集中者及糖尿病者有益。

材料（2 人量）

鮮石斛 15 克

杞子 6 克

鮮鮑魚仔 6 隻

瘦肉 150 克

調味料

海鹽 1/4 茶匙

做法

1. 鮮石斛沖洗，剪成段；杞子浸洗；鮮鮑魚仔去腸臟，洗乾淨；瘦肉切厚片，出水。

2. 全部材料和 1,200 毫升水用中小火煮 1 小時，調味即成。

食療功效

滋補肝腎、益精明目。

飲食宜忌

本品清甜鮮美，補而不燥。適合視力衰退、目昏耳鳴、虛勞精虧、血虛痿黃、腰膝痠軟、糖尿病、脂肪肝患者。痛風患者不宜多食貝類食物。

● **建議每星期服用 1~2 次**

花膠響螺鵪鶉湯

Double-steamed quail soup with fish maw and conch

男：增強精子活力、補腎湯水

材料（2~3 人量）
浸發花膠 150 克
急凍響螺 3 隻
生薑 3 片
紅棗 6 粒
鵪鶉 2 隻

調味料
海鹽半茶匙

做法
1. 花膠、響螺洗淨，出水；紅棗去核；鵪鶉劏洗淨，出水。
2. 將全部材料放入燉盅內，注入開水，隔水燉 3 小時，調味即可連湯料同食。

認識主料

花膠：花膠能補腎益精，滋養筋脈，能治療腎虛滑精。花膠含豐富的蛋白質及膠質，具滋陰養顏的功效，並能補血、補腎、強壯機能，提升性能力；腰膝酸軟、身體虛弱者可經常食用。

鵪鶉：鵪鶉可補五臟、益精血、溫腎助壯。有「動物人參」之稱，屬高蛋白、低脂肪、低膽固醇食物。適合營養不良、體虛乏力、貧血頭暈、腎炎浮腫、高血壓、肥胖者食用。男女經常食用，可增強性功能。

增強男女性能力飲品

簡易飲品助好孕

香港打工一族的工時長、工作繁忙是舉世聞名的，想生育寶寶的在職男女有時想煲些「好孕湯」亦未必有充足時間去購買各種食材，更不用説花時間去烹調有益湯水。但飲品通常較易處理，因為材料較簡單，坊間亦有不少磨成粉末、一沖即可飲用的營養粉劑，如黑豆粉、杏仁粉、黑芝麻粉、核桃粉、紫淮山粉等，更可配合對「助孕」有益的鮮果、奶類等製成各種飲品，提升性能力，增加受孕機會。

不少五穀類、豆類、牛奶、豆漿、小麥胚芽等材料都對促進排卵、健康子宮內膜、增強精子活力各方面有益，有意「備孕」的夫婦不妨多認識各類飲品材料的功效，透過調節全身的基本情況，從而使子宮為受精卵着床及發育提供良好的條件。

黑芝麻核桃露

Black sesame and walnut milk

黑芝麻

龍舌蘭糖漿

認識主料

黑芝麻：黑芝麻含豐富的蛋白質、不飽和脂肪酸、維他命 E 及鈣，鐵、葉酸等。能補血健骨、潤腸通便、防止老化、改善皮膚粗糙，更能提升智力發育，屬好孕食物。維他命 E 能促進卵巢發育完善，使成熟的卵細胞增加。

龍舌蘭糖漿：龍舌蘭糖漿是取自植物龍舌蘭的天然糖分。其所含的皂素不但能防止炎症發生，亦能增強免疫系統及當中的抗菌能力。用龍舌蘭糖漿取代蔗糖於不同食譜中，能減少糖使用的份量，亦同時不會引起血糖的急速上升。

材料（2 人量）
黑芝麻粉、核桃粉各 2 湯匙
牛奶 500 毫升
粟粉水 2 湯匙

調味料
有機龍舌蘭糖漿適量

做法

1. 黑芝麻粉、核桃粉用少許牛奶調成
   稀糊。
2. 牛奶加熱，倒入黑芝麻核桃糊，滾
   起加粟粉水，邊煮邊攪拌，滾起，
   加入龍舌蘭糖漿即成。

食療功效

健腦益智、潤腸通便。

飲食宜忌

本品香滑美味，老少可食。適合習
慣性便秘、皮膚粗糙、血液循環不
好、記憶力衰退、不孕不育者。脾
虛泄瀉者不宜。

● 建議每星期服用 2~3 次

核桃榛子奶露

Walnut and hazelnut milk

材料（2人量）
核桃仁 30 克
榛子仁 30 克
鮮奶 500 毫升
粟粉水 2 湯匙

調味料
砂糖 1 湯匙

做法
1. 核桃仁、榛子仁磨成粉，用開水調成糊。
2. 煮熱鮮奶，加入核桃榛子仁糊、砂糖，待微滾，加入粟粉水，邊煮邊攪拌，略煮片刻即可供服。

認識主料

榛子：堅果是少數含精胺酸的食物之一，有助精蟲的數量和活動力。富含維他命 B_6，可助雌激素和黃體素的平衡。榛子能防治冠心病、動脈粥樣硬化、皮膚癌、宮頸癌等，同時有較強的抗炎作用。

食療功效

補腎固精、潤肺養顏。

飲食宜忌

本品香甜美味，老少可服。適合腎氣不足、腦力衰退、腰痛腳軟、肺弱咳喘、陽痿、遺精、小便頻數者。腹瀉便溏、陰虛火旺者不宜。

● 建議每星期服用 2~3 次

燕麥黑豆漿

Black soymilk with oats

材料（2 人量）

即食燕麥 30 克

黑豆漿 500 毫升

紅棗薄片 1~2 粒份量

調味料

原味砂糖適量

做法

將黑豆漿加熱，加入燕麥及紅棗片煮
5 分鐘，調入砂糖煮溶即可供食。

食療功效

延緩衰老、促進排卵。

飲食宜忌

本品香甜可口，滋潤養顏。適合高
血脂、高血壓、高膽固醇、骨質疏
鬆、不孕不育者。痛風患者不宜。

● 建議每星期服用 2~3 次

認識主料

燕麥：燕麥屬穀類食物，含有大量的維他
命 E、蛋白質、澱粉等，其中維他命 E 不
僅有抗氧化作用，還能夠促進子宮細胞生
長，對子宮發育和卵巢健康都有幫助。

奇亞籽牛奶麥皮

Oatmeal with chia seeds

增強男女性能力飲品

奇亞籽：奇亞籽是 omega-3 脂肪酸豐富的來源，含鈣比牛奶多，含鉀是香蕉的兩倍，含鐵是菠菜三倍之多。常食有助增加精子數量及有助生殖系統的血流量，屬好孕食物。

材料（2 人量）

奇亞籽 1 湯匙
麥皮 30 克
牛奶 500 毫升

調味料

原味砂糖適量

做法

牛奶加熱，放入麥皮及奇亞籽，慢火煮 5 分鐘，加糖調味即可供食。

食療功效

穩定血糖、消脂瘦身。

飲食宜忌

本品香滑美味，老少咸宜。適合三高症、肥胖、水腫、情緒不穩、骨質疏鬆及不孕不育者。一般人群可服。
● 建議每星期服用 2~3 次

杜仲葉茶

Du Zhong tea

增強男女性能力飲品

材料（2人量）

嫩芽杜仲葉 2 湯匙

做法

嫩芽杜仲葉放入壺中，用開水沖洗一遍，再注入開水，焗 5 分鐘即成。

食療功效

補益肝腎、改善性功能。

飲食宜忌

本品味微苦，清香易入口。適合肝腎虧虛、頭暈目眩、腎氣不足、陽痿、遺精者。由於常服杜仲葉茶對體內的糖代謝具有一定的升高作用，故糖尿病患者不宜服。

● 建議每星期服用 2~3 次

認識主料

杜仲葉茶：杜仲葉與杜仲皮具有基本同等功效。能補益肝腎，包含有興奮垂體、腎上腺皮質系統、持續增強腎上腺皮質功能的作用（分泌類固醇荷爾蒙），改善性功能。因此對陽痿、遺精及腎氣不足有較好效果。杜仲葉以嫩芽部分品質最佳。

小麥胚芽茶

Wheat germ tea

材料（2人量）

小麥胚芽粉 2 湯匙

調味料

蜂蜜適量

做法

小麥胚芽粉放入壺內，注入開水拌勻，加蜂蜜調和即可供服。

食療功效

預防三高、增強性能力。

飲食宜忌

本品清香，老少可服。適合血糖高、血脂高、膽固醇高、精力不足者。一般人士可服。

● 建議每星期服用 2~3 次

認識主料

小麥胚芽粉：小麥胚芽是小麥生命的根源，是小麥中營養價值最高的部分。小麥胚芽富含鋅、維他命 B_2、B_6、E，維他命 E 對人類器官的修復及男女的生殖力有很好的影響，能有效促進性荷爾蒙的分泌，可增強性能力。

紅豆紫米水

Red bean and black glutinous rice tea

增強男女性能力飲品

食療功效

滋補肝腎、補血養顏。

飲食宜忌

本品甘甜美味，老少咸宜。適合肥胖、水腫、神疲乏力、貧血、面色蒼白、消化力弱及神經衰弱者。一般人士可服。

● 建議每星期服用 2~3 次

材料（2 人量）
紅豆 30 克
紫米 30 克

調味料
冰糖適量

做法

1. 紅豆、紫米浸洗 2 小時。
2. 用 800 毫升水煮紅豆、紫米 40 分鐘，加入冰糖煮溶即成。
3. 飲湯，隨量吃些湯料。

認識主料

紅豆：紅豆有潤腸通便、降血壓、降血脂、調節血糖、解毒抗癌等功效。婦女在生理期吃些紅豆湯，可以補血、健胃、生津、利尿、消腫、除煩，改善腳氣浮腫。

紫米：紫米有益氣補血、暖胃健脾、滋補肝腎、縮小便、止咳喘等功效。紫米的蛋白質含量及鋅含量較一般稻米高出很多，有助預防男性前列腺肥大、調節前列腺內睾酮的新陳代謝、防止生殖功能障礙、預防神經衰弱、減少膽固醇積聚。

增強男女性能力飲品

蘋果

肉桂粉

認識主料

蘋果：蘋果有補氣健脾、生津、除煩、止瀉等功效。對消化不良、口乾舌燥、妊娠嘔吐等均有幫助。蘋果含鋅量很高，對性腺、腦垂體的發育和活動有幫助，可改善性發育障礙。

肉桂粉：越接近樹幹中心的樹皮所製成的肉桂品質越上等，有發汗、止吐的功效。生理期女性食用肉桂，可緩和不適感；肉桂還可以改善血液循環、手腳冰冷等症狀。對腎陽不足、不孕不育症有幫助。

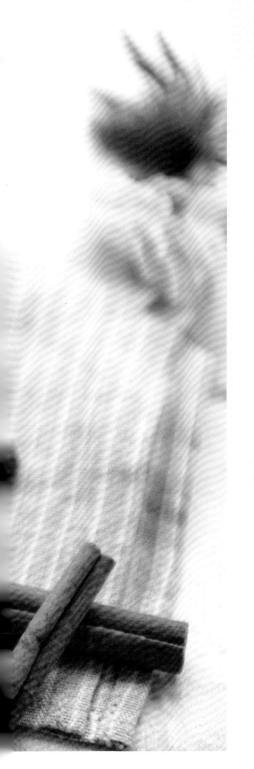

材料（2 人量）

蘋果 2 個

肉桂條 2 條（或肉桂粉半茶匙）

調味料

有機龍舌蘭糖漿適量

做法

1. 蘋果投入開水中，洗去蘋果皮的農藥及蠟質，去核切塊。
2. 將蘋果、肉桂用 400 毫升水煮 15 分鐘，加入糖漿調勻即可供服。

食療功效

健脾益氣、温補腎陽

飲食宜忌

本品芳香濃郁，暖胃祛寒。適合脾胃虛寒、四肢不温、消化不良、血循環欠佳、糖尿病、不孕不育者。內熱上火及孕婦不宜。

● 建議每星期服用 2~3 次

暖宮補血、補腎強精之家常小菜

助孕家常小菜，食得安心又健康

許多打工一族由於工作忙碌，回家已十分疲累，往往無心煮食，隨便去燒臘店買些現成的燒味或滷味，灼碟菜又是一餐；長此以往，飲食營養變得不均衡，身體支出多、收入少，又怎會容易成孕呢！

其實烹調家常小菜，毋須用昂貴材料，或繁複步驟，只要用料新鮮，加上花點心思，發揮一下創意，即可煮到美味窩心的美食，讓你食得安心又健康。

備孕的夫婦，最好多認識各種食材的性味及功效，了解多些有助暖宮補血、補腎強精的餸菜料，例如蠔豉、青口、瑤柱、蝦仁、冬菇、黑木耳、西蘭花、蘆筍、核桃肉、烏雞、乳鴿、花膠等，都屬補腎強體的上好食材，這類食材大多性味平和，補而不燥，食而不膩，只要養好腎臟，健好脾胃，受孕機會自然大大增加。

蠔豉

認識主料

蠔豉：蠔豉能滋陰養血、補益肝腎、活血、生肌。它含有豐富的蛋白質、維他命及礦物質，是補鈣的最好食品。在 10~11 月捕捉、曬乾的蠔豉含糖元及含鋅量特別高，有助護肝及提高男女性能力，同時可增加男性賀爾蒙激素。

蘆筍：蘆筍含葉酸、維他命 A、B、C 及礦物質、植物素等。常食蘆筍能避免生出畸形胎以及治療貧血、提升好孕力，增加懷孕的機率。

材料（2~3 人量）

大隻金蠔 6 隻

紅蘿蔔 1 小段

蘆筍 3 支

五香豆乾 1 塊

冬菇 2 朵

薑汁 2 湯匙

調味料

生抽 1 湯匙

蠔油 2 茶匙

海鹽半茶匙

米酒適量

做法

1. 金蠔浸洗淨，蒸半小時，切粗粒；紅蘿蔔去皮，切粗粒；蘆筍洗淨，切粒；五香豆乾切粗粒；冬菇浸軟，去蒂後切粒。

2. 燒熱少許油，放入蠔粒煎香，再加入其他材料，潷酒，加入薑汁及調味料，加少許水，煮大約 10 分鐘至汁乾即成。

食療功效

補益五臟、提高性功能。

飲食宜忌

本品香濃味美，老少可食。適合陰虛煩熱、氣血不足、精神倦怠、貧血及不孕不育者。脾胃虛寒及痛風患者不宜。

● **建議每星期服用 1~2 次**

乾貝絲燴西蘭花

Broccoli in dried scallop glaze

暖宮補血、補腎強精家常小菜

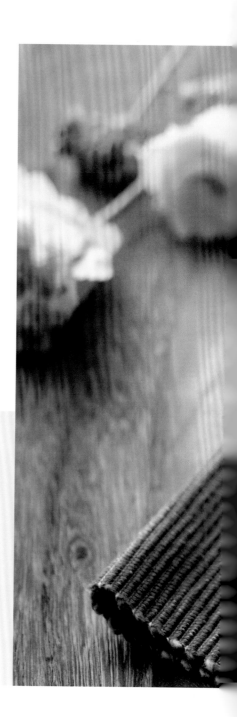

食療功效

消脂降壓、助孕強精。

飲食宜忌

本品鮮甜美味、滋陰補虛。適合三高症、糖尿病、肥胖、食欲不振、不易受孕者。一般人群可食。

● 建議每星期服用 1~2 次

暖宮補血、補腎強精家常小菜

瑤柱：瑤柱又稱乾貝，能消脂降壓、消食、抑制腫瘤、養陰補虛。適合脾胃虛弱，中氣不足、營養不良，或久病體虛者食用。

材料（2~3人量）
蒸過瑤柱 4 粒（拆絲）
西蘭花 1 棵
薑蓉 1 茶匙
高湯 100 毫升
生粉水 2 湯匙

調味料
海鹽半茶匙
蠔油 1 湯匙

做法
1. 西蘭花洗淨，切塊後放入加了油、鹽的沸水中煮熟，撈出排放碟中。
2. 燒熱少許油，爆香薑蓉，加入蒸軟的瑤柱絲、高湯及調味料，煮至滾起，加入生粉水埋薄芡，連汁料淋在西蘭花上即成。

西蘭花：西蘭花具有抗癌及提高免疫力的作用，可以幫助身體吸收營養、調養氣血，防骨質疏鬆，更有助促進卵子的滋潤，亦有助產生健康的精子。

暖宮補血、補腎強精家常小菜

彩椒核桃雞丁

暖宮補血、補腎強精家常小菜

琥珀核桃

西椒

認識主料

琥珀核桃：琥珀核桃是以核桃肉加工製成的傳統小食，琥珀色，口味香脆甘甜。有補腎固精、溫肺定喘的功效，對腎陽虛氣弱、腦力衰退、陽痿、遺精、小便頻數、腰痠腿弱、咳嗽氣喘有益。

西椒：西椒與辣椒相比，辣味較淡，具有較高的含糖量和維他命，更富含蛋白質及無機礦物質中的鈣、鈉、磷、鐵及維他命A、C、菸酸等，還含有指甲和毛髮生長所需營養的矽元素，對肌膚有活化細胞組織功能，促進新陳代謝，令皮膚光澤細嫩。所含纖維多，有助排泄有害物質，增加食欲，對眼睛亦非常有益。

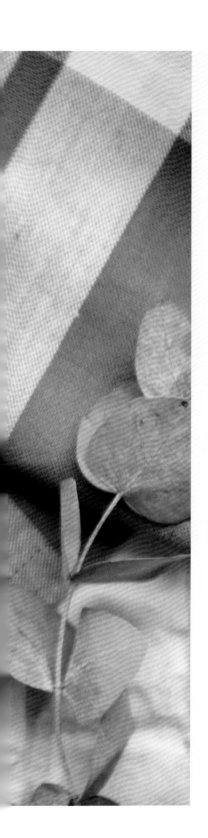

材料（2~3 人量）　醃肉料

彩色甜椒 150 克　　生抽 2 茶匙

琥珀核桃 100 克　　米酒 2 茶匙

雞腿肉 200 克　　　鹽 1/4 茶匙

薑蓉 2 茶匙　　　　生粉及胡椒粉適量

調味料

海鹽半茶匙

米酒適量

做法

1. 彩椒洗淨，去核切丁；雞腿肉洗淨，切丁後用醃料醃半小時。

2. 燒熱少許油，爆香薑蓉，加入雞丁炒香，加彩椒丁兜炒，灒酒加調味料，最後加入琥珀核桃炒勻即可上碟。

| 食療功效 |

溫中益氣、補腎固精。

| 飲食宜忌 |

本品甘香美味、滋補強身。適合虛勞瘦弱、食欲不振、腦力衰退、腰膝痠軟、陽痿、遺精者。一般人群可食。

● 建議每星期食用 1~2 次

韭菜蝦仁炒蛋

Scrambled egg with Chinese chives and shrimps

暖宮補血、補腎強精家常小菜

材料（2~3 人量）

韭菜 60 克
鮮蝦仁 100 克
雞蛋 2 個

調味料

海鹽半茶匙
油 1 湯匙
胡椒粉適量

做法

1. 韭菜去葉尾枯黃部分，洗淨切碎；蝦仁去腸，沖洗乾淨後瀝乾。
2. 雞蛋打散，加調味料打勻。
3. 燒熱 2 湯匙油，先將蝦仁煎香，加韭菜略炒，倒入蛋漿，炒至剛熟就成。

認識主料

蝦仁：蝦肉能補氣健胃、暖腎壯陽。對脾胃虛弱、神經衰弱、腎虛陽痿、腰膝痠軟、神疲乏力者有益。蝦含鋅量高，能增加精子活力及提升性功能。

食療功效

溫中健胃、補腎壯陽

飲食宜忌

本品清香味美、滋補強壯。適合精神不振、神疲乏力、腰膝痠軟、腎虛陽痿、月經不調者。一般人群可食，痛風患者宜少食。

● 建議每星期服用 1~2 次

潮式煎蠔烙

Oyster omelette, Chuichow style

暖宮補血、補腎強精家常小菜

食療功效

滋陰養血、強壯體質。

飲食宜忌

本品外酥內軟,滋味可口。適合脾胃虛弱、貧血、精神不振、骨質疏鬆者。但氣滯、脂肪肝、痛風患者宜少食。

● 建議每星期服用 1~2 次

材料（2~3 人量）
蠔仔 250 克
鴨蛋 2 個
蔥花 1 湯匙
芫茜碎 1 棵份量
白胡椒粉適量

粉漿
番薯粉 2 湯匙
粟粉 1 湯匙
水半杯

調味料
鹽 1 茶匙

做法

1. 蠔仔用生粉、鹽搓洗乾淨，瀝乾；鴨蛋打成蛋液。

2. 粉漿加入蔥花及調味料拌勻，再加入蠔仔。

3. 燒熱 4 湯匙油，先倒入蠔仔粉漿煎片刻，再加入蛋液，煎至兩面金黃，灑入芫茜碎及胡椒粉即成。

暖宮補血、補腎強精家常小菜

蠔仔：蠔仔即蚵仔，主要功效有滋陰養血、軟堅散結、收斂固澀、止虛汗等作用。蠔仔和蠔肉一樣，含有豐富的荷爾蒙，可補血兼提高男性的性能力，治療男性陽痿，並對胃痛、失眠、甲狀腺功能減退及甲狀腺腫等疾病有一定的輔助治療作用

鴨蛋：鴨蛋能滋陰養血、生津益胃。其所含的營養可與雞蛋媲美，鴨蛋中的蛋白質含量和雞蛋相當，而礦物質總量遠勝雞蛋，尤其鐵、鈣含量極豐富，能預防貧血，促進骨骼發育。

韓式蘿蔔燜牛肋骨

Korean braised beef ribs with radish

● 暖宮補血、補腎強精家常小菜 ●

認識主料

牛肋條：牛肋條是牛肉近肋骨的肉，牛肉能補中益氣、滋養脾胃，並含豐富蛋白質、鐵和鋅，能提高人體抗病及性功能，和有補血、強筋健骨功效。

食療功效

補益脾胃、強壯筋骨。

飲食宜忌

本品香味濃郁、滋補強壯。適合氣血虛弱、營養不良、貧血、面色痿黃、筋骨痿軟、性欲衰退者。一般人群可食，有皮膚舊患者宜少食。

● **建議每星期食用 1~2 次**

材料（3 人量）

牛肋骨 400 克

白蘿蔔 200 克

紅椒 1 隻

蒜頭 2 粒

生薑 3 片

芫茜碎 1 湯匙

調味料

韓式燒牛肉汁半瓶

鹽半茶匙

米酒 1 湯匙

做法

1. 牛肋骨出水，用少許油兩面煎香；白
   蘿蔔去皮，切塊；紅椒去核，切丁。

2. 將煎香的牛肋骨及其他材料（除芫茜
   碎外）放入鑊內，加調味料及清水蓋
   過料面，大火滾起，改用中火煮 1 小
   時，待汁收至濃稠，灑入芫茜碎即成。

花膠扒時蔬

Braised fish maw with leafy greens

暖宮補血、補腎強精家常小菜

材料（2~3 人量）
細隻浸發花膠筒 6 隻
小棠菜 6 棵
生薑 3 片
葱白 3 條

調味料
蠔油 2 湯匙
清雞湯 150 毫升
米酒適量
生粉水 1 湯匙

做法

1. 浸發花膠放入有生薑、葱白的水內出水，撈出備用。
2. 小棠菜洗淨，剖開一半，放入滾水中燙至熟，排放碟中、
3. 煮滾雞湯，放入花膠煮 10 分鐘至夠腍，排放碟中；把蠔油及米酒加入雞湯內，滾起，埋薄芡，淋在菜面即成

食療功效

補腎益精、增強性能力。

飲食宜忌

本品香滑美味、滋補有益。適合身體虛弱、精神不振、神疲乏力、腰痠背痛、性欲減退者。一般人群可食，外感發燒不宜。
● 建議每星期服用 1~2 次

認識主料

花膠介紹請看第 66 頁

暖宮補血、補腎強精家常小菜

甜茴香乳鴿

Braised squab in fennel sauce

暖宮補血、補腎強精家常小菜

食療功效

温腎散寒、提高性功能。

飲食宜忌

本品清香味美，滋補強身；適合脾胃消化力弱、宮寒不孕、陽痿、遺精不育者。陰虛火旺及孕婦不宜。

● 建議每星期食用 1~2 次

材料（2 人量）

乳鴿 1 隻
甜茴香 1 棵
杞子 5 克
薑 2 片

調味料

滷水汁 2 湯匙
生抽 1 湯匙
海鹽 1/4 茶匙
水晶冰糖 4 粒

做法

1. 甜茴香洗淨，切片；杞子沖淨，乳鴿整隻劏洗淨。
2. 燒熱少許油，爆香薑片、茴香片，乳鴿下鑊稍煎，加入杞子、調味料及 200 毫升水，慢火煮 15 分鐘，上碟時將乳鴿斬件排放，淋上汁料即成。

認識主料

甜茴香：甜茴香又名結球茴香，可促進食欲，解胸悶，助消化，溫腎散寒。婦女授乳期間，促進母乳大量分泌。茴香植物含類似人體雌激素成分，可調節荷爾蒙與內分泌平衡。

乳鴿介紹請看第 29 頁

三杯雞

Three-cup chicken

暖宮補血、補腎強精家常小菜

認識主料

羅勒葉介紹請看第 26 頁

雞肉：雞肉温中益氣、補精添髓。含豐富的蛋白質及不飽和脂肪酸。對心血管患者及體質虛弱、病後或產後人士最適合。對虛勞瘦弱、骨蒸潮熱、脾虛泄瀉、婦女崩漏、白帶及男子遺精等都有益。

材料（2~3 人量）

羅勒葉 30 克
嫩雞 1 隻
蒜頭 8 瓣
紅椒 2 隻
生薑 20 克

調味料

麻油 1 小杯
醬油膏 1 小杯
米酒 1 小杯
糖 2 茶匙

做法

1. 嫩雞劏洗淨，斬件；羅勒葉去枝，洗淨；生薑切片；蒜頭去衣；紅椒切片。

2. 燒熱麻油，將雞件煎至金黃，盛起。

3. 用鑊中餘油爆香薑片、蒜頭及紅椒，倒入雞塊，灒酒，加入其餘調味料，轉中慢火煮約 10 分鐘至汁乾，灑入羅勒葉，兜炒片刻即可上碟。

食療功效

健脾暖胃、補精強身。

飲食宜忌

本品香濃美味、滋補強身。適合脾胃虛寒、食欲欠佳、精神不振、腰膝無力及不易受孕者。一般人士可食。外感發熱及陰虛火旺者少食。

● **建議每星期食用 1~2 次**

暖宮補血、補腎強精家常小菜

蒜蓉豉椒蒸青口

Steamed mussels in garlic black bean sauce

暖宮補血、補腎強精家常小菜

認識主料

青口：青口曬乾亦即煲湯常用的淡菜，能補肝腎、益精血、消癭瘤。對男性陽痿遺精、腰痛，女性崩漏、白帶及盜汗、眩暈等症均有幫助，屬好孕食物。男子常食可強壯身體增強性功能、提高精子的質量。

材料（2~3 人量）
急凍青口肉 10 隻
豆豉 2 茶匙
蒜蓉 2 茶匙
薑蓉 2 茶匙
紅椒碎 2 茶匙
芫茜碎 2 茶匙（或免）

調味料
熟油、生抽適量

做法

1. 青口解凍，洗淨瀝乾，排蒸碟中；豆豉剁碎，與蒜蓉、薑蓉、紅椒碎拌勻，放在每隻青口肉上，隔水大火蒸 7 分鐘。
2. 灑入芫茜碎，淋上滾油及生抽即可供食。

食療功效

滋補肝腎、益血填精。

飲食宜忌

本品香濃美味，老少可食。適合虛勞瘦弱、貧血、眩暈、盜汗、陽痿、遺精、白帶、宮冷不孕者。一般人群可食，痛風患者宜少食。

● **建議每星期服用 1~2 次**

暖宮補血、補腎強精家常小菜

促進生育之主食

讓人垂涎、促進生育的粥粉麵飯

粥粉麵飯是餐桌上的主食，是人體所需能量的主要來源。中國人的主食以米飯為主，但近年受到西方飲食文化影響，麵食、通粉、意粉等越來越受歡迎。

主食提供了人們所需的大部分卡路里，一般以澱粉為主要成分，但不同的天然穀物可以製作出各種令人垂涎的美食，材料可謂變化多端，同樣是一碗飯、一碗粥、一碗麵食，只要有計劃地採用對精子、卵子有益的食材作配料，也可以烹調出變化萬千的精緻主食，讓備孕的男女吃得多、吃得高興，增加成孕機會。

牛油果蓉椰菜仔螺絲粉

Fusilli with Brussels sprouts in avocado beef sauce

牛油果

初榨牛油果油

認識主料

牛油果：牛油果又稱酪梨，是一種營養非常豐富的果實，牛油果是維他命 E 的優良來源，有助於強化子宮內膜以及胚胎着床。女性經常食用，有助平衡雌激素，防宮頸癌。

初榨牛油果油：初榨牛油果油含較高不飽和脂肪酸，帶堅果的香氣，含植物性奧米加 3、6、9，能降膽固醇，預防心血管病，所含天然葉酸對促進生長發育，對精子及卵子健康均甚為有益。

117

促進生育之主食

材料（2 人量）　醃肉料

牛油果 1 個　　　生抽 2 茶匙

初榨牛油果油 2 湯匙　米酒 2 茶匙

椰菜仔 6 粒　　　粟粉 1 茶匙

鮮蘑菇 4 粒

絞碎牛肉 50 克　調味料

螺絲粉 150 克　　海鹽，胡椒粉適量

做法

1. 牛油果剖開，去核，將肉刮出壓成蓉；椰菜
仔洗淨，切開對半，蘑菇去蒂，切片。椰菜
和蘑菇放入開水中燙片刻撈出；絞碎牛肉用
醃料醃至入味。

2. 螺絲粉放入開水中煮 12 分鐘，撈出放碟中。

3. 將牛油果油倒入平底鑊煮熱，加入碎牛肉煎
香，再加入椰菜仔、蘑菇片及牛油果蓉，潷
酒，加入半碗水及調味料，煮 5 分鐘，淋在
螺絲粉上即可供食。

食療功效

健脾和胃、消除疲勞。

飲食宜忌

本品美味可口，老少可食。適合食欲欠佳、
容易疲倦、精神不振，不易受孕者。一般人
士可服。

● 建議每星期服用 2~3 次

韭菜肉末餃子

● 促進生育之主食 ●

食療功效

溫中健胃、壯陽固精。

飲食宜忌

本品香甜美味，老少可食。適合陽虛怕冷、四肢不溫、腰膝痠軟、遺精、宮冷者。一般人士可食。陰虛火旺者不宜。

● 建議每星期服用 1~2 次

材料（2~3 人量）

韭菜 100 克

絞碎豬肉 300 克

蛋白 1 個

薑蓉 1 茶匙

餃子皮 200 克

調味料

海鹽半茶匙

雞粉 1 茶匙

糖半茶匙

米酒 1 湯匙

粟粉 1 湯匙

蘸料

切幼薑絲 2 茶匙

麻油、餃子醋適量

做法

1. 韭菜洗淨，去老葉後切碎，加入少許油拌勻以免溢出水分；將韭菜、絞碎豬肉加入蛋白、薑蓉及調味料，攪拌至起膠。

2. 將攪拌好的餡料用餃子皮包好，投入大滾水中，煮至浮起，加入半碗清水，再次浮起，即可撈出蘸醬料供食。

認識主料

韭菜：韭菜為振奮性強壯食物，有健胃、提神、溫暖作用。適用於肝腎陰虛、尿頻、陽痿、遺精，婦女月經病、經痛、經漏、帶下等症。

椰香薑黃雞粒藜麥飯

Coconut-scented quinoa rice with chicken and turmeric

薑黃介紹請看第 157 頁

藜麥

認識主料

藜麥：藜麥含有較高的植物雌激素，這種物質對預防乳腺癌、骨質疏鬆及改善女性的經前症候群與停經後的更年期症狀很有效。所含豐富的礦物質可以緩解血管壓力，減少心臟病的發生，對調節血糖及降低膽固醇亦有功效。

材料（2~3 人量）

雞柳 200 克

蘆筍 3 支

藜麥 30 克

白米 50 克

薑黃粉 3 茶匙

椰漿 5 湯匙

蒜頭 3 瓣

醃肉料

米酒 2 茶匙

海鹽 1/4 茶匙

胡椒粉、粟粉各少許

調味料

海鹽適量

做法

1. 雞柳洗淨，切粗粒，用醃料醃半小時；蘆筍洗淨，切粒；藜麥用暖水浸 2 小時，白米沖洗。

2. 藜麥、白米加適量水放電飯煲加熱煮好；蒜頭去衣，剁碎。

3. 燒熱少許油，爆香蒜蓉，加入雞肉粒、蘆筍粒炒香；加入藜麥飯，再加入薑黃粉兜炒，倒入椰漿及調味料，炒勻即可供食。

食療功效

健脾益氣、消脂瘦身。

飲食宜忌

本品香味濃郁、滋補強身。適合高血脂、高血壓、高膽固醇、肥胖、食欲不振、記憶力減退、月經症候群者。陰虛火旺者及孕婦不宜用薑黃粉。

● 建議每星期食用 1~2 次

韓式鰻魚飯

Korean grilled eel rice

食療功效

滋補強身、增強性能力。

飲食宜忌

本品甘香美味、滋補強壯。適合虛勞瘦弱、貧血、頭暈心悸、精神不振、風濕痹痛及性欲減退者。一般人士可食,痰濕盛者宜少食。

● 建議每星期食用 2~3 次

材料（2 人量）

燒鰻魚 1 條

甜酸子薑絲 1 小撮

蛋皮絲 1 小撮

炒香白芝麻 1 小撮

牛油生菜 1 棵

白飯兩碗

做法

1. 牛油生菜逐片葉片洗淨，瀝乾水分；燒鰻魚切塊。

2. 取一大片牛油生菜，將白飯、燒鰻魚、甜酸子薑絲、蛋皮絲及炒香芝麻各適量包好即可供食。（也可以用西生菜代替牛油生菜）

認識主料

鰻魚：鰻魚滋補強身，為營養價值極高的蛋白質食品，含豐富維他命及礦物質，鰻魚有助降低膽固醇及預防血管疾病的作用，鰻魚體內有一種稀有的西河洛克蛋白，能增強男性的生殖能力。

牛油生菜：牛油生菜葉片翠綠，肥厚爽脆，最好買到不用泥土種植的水耕牛油生菜，吃起來沒有泥味，清洗亦容易，不怕殘餘農藥和重金屬。它含豐富的纖維及維他命 C，有降低膽固醇、利尿、促進血液循環、抗病毒等作用。

帶子薑蛋炒飯

認識主料

帶子：帶子營養豐富，屬高蛋白、低脂肪食物，容易消化，並對降低血清膽固醇有獨特作用。帶子很適合脾胃虛弱、營養不良或久病體虛，脾腎陽虛及夜尿多者食用。

材料（2人量）　　　醃帶子料

急凍帶子 150 克　　雞粉 2 茶匙
雞蛋 2 個　　　　　米酒及粟粉各少許
急凍雜菜粒 100 克
薑蓉 2 茶匙　　　　調味料
冷飯 2 碗　　　　　海鹽半茶匙
　　　　　　　　　胡椒粉少許

做法

1. 急凍帶子解凍，沖洗後用醃料醃半小時，投入開水中，迅即熄火浸至八成熟，撈出備用。
2. 雜菜粒解凍，出水；雞蛋打散。
3. 燒熱少許油，加入一半蛋漿，倒入打散了的冷飯、薑蓉、雜菜粒及調味料，兜炒均勻後，加入帶子及餘下蛋漿，快手炒勻即可上碟供食。

食療功效

滋補肝腎、健腦益智。

飲食宜忌

本品清香味美，老少可食。適合營養不良、食欲欠佳、腦力衰退、夜尿多者。一般人士可食，痛風者宜少食帶子。

● 建議每星期食用 1~2 次

紫淮山圓肉百合粥

Purple yam congee with dried longans and lily bulbs

認識主料

百合：百合能潤肺止咳、寧心安神、補中益氣、清熱利尿、美容養顏、防癌抗癌。百合含植物雌激素，對卵巢有保護作用，且有防癌功效，對虛煩失眠人士有益。

材料（3 人量）

鮮紫淮山 100 克
圓肉、百合各 30 克
白米 60 克

調味料

冰糖適量

做法

1. 紫淮山去皮洗淨，切成粗粒。
2. 百合浸洗；圓肉沖淨；白米洗淨。
3. 全部材料放入電飯煲內，加適量水煮成濃稠適度的粥，加入冰糖，待糖溶即可供食。

食療功效

健脾補腎、養心安神。

飲食宜忌

本品香滑甜美，老少咸宜。適合貧血、容易疲倦、失眠、腹瀉、子宮下垂、遺精及不孕不育者。一般人士可食。濕熱體質及便秘者不宜。

● **建議每星期服用 1~2 次**

南瓜子杞子小米粥

Millet congee with pumpkin seeds
and goji berries

促進生育之主食

材料（2 人量）

南瓜子 20 克

杞子 6 克

小米 60 克

藕粉 1 湯匙

調味料

紅糖適量

做法

1. 杞子浸洗，藕粉用少許涼開水調成糊狀。

2. 小米浸洗後，連同南瓜子用 6 碗水煮半小時，加入杞子、紅糖及藕粉糊，邊煮邊攪，煮至黏稠即成。

食療功效

滋補肝腎、提升性功能。

飲食宜忌

本品香滑甜美，老少可食。適合脾胃虛弱、貧血、營養不良、煩躁口渴、月經量多者。一般人士可食。

● 建議每星期食用 1~2 次

認識主料

南瓜子：南瓜子含有非常豐富的營養物質：鐵、鋅、鎂、錳等，由於鋅含量很高，有助改善精子品質及提升性功能，並能保護男性前列腺，消除前列腺發炎及預防前列腺癌。

藕粉：富含澱粉、鐵、鈣、氨基酸、維他命 B_{12}、維他命 C 等，既易消化，又能養胃滋陰、養血止血；同時藕粉能幫助睡眠，增強人體免疫功能。

蚵仔粥

Baby oyster congee

促進生育之主食

食療功效

滋陰養血、提高性功能。

飲食宜忌

本品鮮甜美味、滋補強身。適合
體質虛弱、食欲欠佳、精神疲倦、
失眠、甲狀腺病及性欲減退者。
一般人士可服，痛風患者不宜。
● 建議每星期服用 1~2 次

認識主料

蠔仔介紹請看第 102 頁

材料（2 人量）
蠔仔（蚵仔）150 克
鮮魚湯 500 毫升
冬菇 2 朵
絞碎豬肉 50 克
葱絲、薑絲各 1 湯匙
白米 60 克

調味料
海鹽、胡椒粉適量

做法
1. 蠔仔用生粉搓揉洗淨，出水後撈出備用。
2. 白米洗淨，用少許油、鹽略醃；冬菇浸軟去蒂，切幼絲。
3. 白米、冬菇絲、薑絲、鮮魚湯及適量水放電飯煲內，煮成濃稠適度的粥。
4. 將肉碎、蠔仔放入粥內煮 5 分鐘，加調味料，最後灑入葱絲即可供食。

增強卵子、精子活力飲品甜點

美味飲品、甜點，也可起到助孕功效

一般人的觀念，認為女性偏好甜點；但有研究顯示，不論男性、女性，都有可能對甜食上癮，皆因吃了甜食之後，身體當中的血糖含量會增高，胰島素會將其轉化為身體能量，吃甜食越多，身體的能量就越充足，人就會產生很快樂的感覺。

因此，要達成好孕的目的，有意生育的夫婦，不妨在休閒的日子，精心策劃製作一些消除壓力、舒緩情緒、幫助卵子、精子健康活力的飲品及甜點，只要限制用糖量，減少過量奶油、麵粉，製作些不易肥胖及不影響健康的輕甜食，夫婦在共享濃情輕甜食之餘，成孕機會亦會大增。

黑豆漿

認識主料

薑汁：薑汁能散寒、止嘔、解表，可促進食欲、緩解經痛及扭傷腫痛、預防感冒、止暈止嘔等。生薑汁溫經散寒功效頗佳，對婦女宮冷瘀血多很有幫助。

黑豆漿：黑豆含豐富的優質蛋白、維他命 B 群及維他命 E，還有鐵、鈣、磷、鉀、鎂及花青素等，可以養顏抗拒衰老，並有保護子宮正常發育的功效。女性宜多飲黑豆漿，能促進排卵，令子宮壁增厚，有利於懷孕。

材料（2 人量）

薑汁 2 湯匙

黑豆漿 250 毫升

蛋白 3 個

調味料

砂糖 30 克

做法

1. 砂糖用少許暖水調至完全溶化，加入蛋白、薑汁、黑豆漿調至均勻。

2. 倒入深碟中，撇去表面少許泡沫，用鋁箔紙把碟包住，隔水蒸 12 分鐘即可供食。

食療功效

暖胃健脾、促進排卵。

飲食宜忌

本品香滑甜美，滋補養顏。適合宮寒不孕、子宮壁薄弱、營養不良、不孕不育者服食。一般人士可服。

● 建議每星期服用 1~2 次

酒釀竹絲雞蛋糖水

Poached Egg sweet soup with distiller's grains

增強卵子、精子活力飲品甜點

材料（1人量）
酒釀 100 克
竹絲雞蛋 1 個

調味料
原味砂糖 2 茶匙

做法

1. 酒釀加清水 1 杯 (250 毫升)
   煮滾，加入砂糖及打入一個
   竹絲雞蛋，熄火焗 2 分鐘。

2. 再按自己喜好蛋的熟度，慢
   火煮片刻即供食。

認識主料

竹絲雞 / 蛋：竹絲雞和蛋有滋
陰潤燥、補益肝腎功效。對氣
血不足、熱病煩渴、骨蒸羸瘦、
月經不調、胎動不安及不育症
者都有裨益。

酒釀：酒釀能促進血液循環、
促進新陳代謝，具有補血養
顏、舒筋活絡、強身健體和延
年益壽的功效。酒釀還有促進
食欲、幫助消化、溫寒補虛、
舒緩神經衰弱、精神抑鬱、健
忘等症狀。

食療功效

益氣生津、滋陰補虛。

飲食宜忌

本品酒香撲鼻，補而不燥。對氣血不足、食欲不振、月經不調、體虛不育者有益，常服有助改善體內的腸道環境，有美容、豐胸功效。一般人士可服。

● 建議每星期服用 1~2 次

玫瑰合歡花茶

Rose tea with He Huan Hua

● 增強卵子、精子活力飲品甜點 ●

食療功效

舒緩情緒、平衡內分泌。

飲食宜忌

本品清香，養顏護膚。適合精神緊張、抑鬱、失眠、月經不調、宮冷不育者。但孕婦不宜飲花草茶。

● 建議每星期食用 2~3 次

材料（1 人量）

玫瑰花 1 湯匙

合歡花蕾 1 湯匙

做法

玫瑰花、合歡花放入壺內，用開水沖洗一次，再注入開水，焗 7 分鐘即可飲用。

認識主料

玫瑰：玫瑰可緩和情緒、平衡內分泌、補血氣，美顏護膚。常服玫瑰花茶，能鎮定經前緊張症狀，促進陰道分泌，調節月經週期。對不孕症、性欲低下會有幫助。

合歡花：合歡花乾燥花序呈團塊狀，有如棉絮。合歡米是合歡花的乾燥花蕾。無論合歡花、合歡米皆能解鬱安神，和絡止痛。對肝鬱胸悶，憂鬱不樂，健忘失眠者頗為有益。

艾葉紅糖水

增強卵子、精子活力飲品甜點

艾葉

紅糖

認識主料

艾葉：艾葉有溫經、散寒、止血、消炎、平喘、止咳、安胎、抗過敏等作用。艾葉對宮寒不調或宮冷不孕症很有幫助。

紅糖：紅糖具有益氣補血、健脾暖胃、緩中止痛、活血化瘀的作用。紅糖含有多量的葡萄糖和胡蘿蔔素、核黃素、菸酸和多種微量元素，對於子宮復元有幫助。

材料（1 人量）
艾葉 10 克

調味料
紅糖 1 湯匙

做法
艾葉洗淨，用 4 碗水煮 20 分鐘，加入紅糖
煮溶即可供服。

食療功效

祛寒暖身、調理月經。

飲食宜忌

本品味道微苦，適合宮寒不調及宮冷不孕
症。對孕婦更有暖宮止血安胎作用。較適
合女性服用；陰虛火旺、血熱病及失血症
者忌用。
● 建議每星期食用 1~2 次

百合茯苓茶

Fu Ling and lily bulb tea

增強卵子、精子活力飲品甜點

認識主料

茯苓：茯苓能利水滲濕、健脾安神。對水腫、脾虛食少、便溏泄瀉、心神不安、失眠者有益。茯苓含有多醣類物質，對預防卵巢囊腫很有幫助。

材料（1 人量）
百合、茯苓各適量
杞子 2 克

調味料
砂糖適量

做法

1. 百合、茯苓磨成粉；杞子沖淨。
2. 燒熱 250 毫升水，加入百合、茯苓粉各 1 湯匙及杞子，煮至滾起，調入砂糖即可供服。

食療功效

寧心安神、保護卵巢。

飲食宜忌

本品甘香，簡單易煮，無論男女飲用，都有助清熱除煩、安靜神經。適合虛煩不安、睡眠不好、不易受孕者飲用。一般人士可服，但虛寒精滑、氣虛者慎服。

● 建議每星期服用 1~2 次

山楂荔枝乾糖水

Haw and dried lychee sweet soup

材料（2人量）

山楂 10 克
荔枝乾 10 粒
杞子 6 克
紅棗 5 粒

調味料

紅糖適量

做法

1. 將山楂、杞子沖洗；荔枝乾去殼取肉；紅棗去核。
2. 全部材料用 4 碗水煮 20 分鐘，調入紅糖煮溶，即可連湯料同食。

認識主料

山楂：山楂能開胃消食、化滯消積、活血散瘀、化痰行氣；還可以調節月經、痛經；並有增強免疫力、抗衰老及抗癌功效。

荔枝乾：荔枝乾具有生津止渴、補氣養血、理氣止痛等功效。尤其對病後體虛、婦女貧血、月經過多、胃脘脹痛、胃陰不足及口渴咽乾等症有益。

食療功效

健脾補血、活血散瘀。

飲食宜忌

本品甜酸可口，補血養顏。適合貧血、月經不調、痛經及宮寒不孕婦女服用。對男性疝氣、睪丸腫痛等症亦有幫助，男女可服。但陰虛火旺及孕婦不宜。

● 建議每星期服用 1~2 次

薑黃蜜奶

Honey milk with turmeric

材料（1 人量）
薑黃粉 1 茶匙
鮮奶 250 毫升

調味料
蜂蜜適量

做法
鮮奶慢火加熱，毋須滾起即可加入薑黃粉調勻，微滾熄火。飲時加蜂蜜即成。

食療功效

健胃利膽、調經助孕。

飲食宜忌

本品金黃色，香甜美味，男女皆可飲用。睡前服用最佳。適合月經不調、關節疼痛、血液循環及睡眠質素欠佳人士服用，常服更有助孕及預防腦退化的作用。但有膽結石、懷孕及陰虛火旺、口乾舌燥者不宜。

● 建議每星期服用 1~2 次

認識主料

薑黃：薑黃有促進膽汁分泌、調經止痛等功效。所含薑黃素成分有抗氧化、抗發炎、利膽、加強腸胃機能、幫助肝臟解毒等食療作用。對於改善月經遲來、月經失調和閉經症狀有一定的益處。

牛奶：牛奶營養豐富，具有補虛損、益肺胃、生津潤腸的功效。牛奶同時能提升免疫力，助安睡，改善大腦功能，促使卵子更有活力，增加成孕的機率。

牛油果香蕉奶昔

Avocado banana smoothie

增強卵子、精子活力飲品甜點

材料（3 人量）

牛油果 2 個

香蕉 2 條

鮮奶 500 毫升

鮮檸汁 1 茶匙

做法

1. 牛油果去核，將肉刮出；香蕉去皮，切塊。
2. 將牛油果肉、蕉肉放入攪拌機內攪成蓉，再加入鮮奶打至幼滑，滴入鮮檸汁即可飲用。

食療功效

補益肺胃、強壯體質。

飲食宜忌

本品香滑美味，滋補養顏。對貧血、高血壓、胃酸過多、習慣性便秘及不孕不育者有益。脾胃虛寒及腹瀉者忌服。

● 建議每星期服用 2~3 次

認識主料

牛油果介紹請看第 117 頁

香蕉：香蕉有清熱、生津止渴、潤肺滑腸的功效。香蕉的維他命 B 含量高，可抗抑鬱，含糖類物質可充饑及補充營養及能量；所含多種礦物質對貧血、胃潰瘍、高血壓等均有幫助。

増強卵子、精子活力飲品甜點

Pork tripe soup with white peppercorns and black beans

(makes 3 to 4 servings) Ref.p.019

Ingredients

5 g white peppercorns
30 g black beans with green kernels
1 pork tripe
200 g lean pork
1 sprig coriander (finely chopped)

Seasoning

1/2 tsp sea salt

Method

1 Rinse the black beans. Drain and set aside. Rub coarse salt and caltrop starch on the pork tripe repeatedly. Rinse well after rubbing. Blanch pork tripe in boiling water. Drain and cut into pieces. Set aside. Cut lean pork into chunks. Blanch in boiling water and drain.

2 Put all ingredients into a pot. Add 1.5 litres of cold water. Bring to the boil over high heat. Turn to medium heat and cook for 2 hours. Season with sea salt and sprinkle with chopped coriander. Serve.

Indications and contraindications

This aromatic soup whets the appetite and is suitable for those with physical exhaustion, skinny build, poor appetite, Qi- and Blood-Asthenia, excessive leucorrhoea, loose stool and diarrhoea. However, those with excessive phlegm-Dampness, Yin-Asthenia with overwhelming Fire, or those suffering from cold or flu with fever should avoid.

Suggested dosage

Serve once or twice each month.

Pork rib soup with papaya and peanuts

makes 3 to 4 servings) Ref.p.022

Ingredients

1 green papaya
30 g peanuts
1 whole dried tangerine peel
4 red dates
300 g pork ribs

Seasoning

1/2 tsp sea salt

Method

1 Peel and rinse the papaya. Cut into chunks. Set aside. Soak and rinse peanuts and dried tangerine peel. Drain and set aside. De-seed the red dates. Blanch pork ribs in boiling water. Drain.
2 Put all ingredients into a pot. Add 1.5 litres of water. Bring to the boil over high heat. Turn to medium-low heat and cook for 1 hour. Season with sea salt. Serve.

Indications and contraindications

This soup is delicious and is good for all ages. It is suitable for those with Spleen- and Stomach-Asthenia, general weakness after recovery from sickness, malnutrition, poor appetite or hormonal deficiency. Generally speaking, everyone may consume, except pregnant women who should avoid.

Suggested dosage

Serve once or twice each week.

Lean pork soup with beetroot and cashews

(makes 2 to 3 servings) Ref.p.024

Ingredients

1 beetroot
2 tomatoes
30 g cashew nuts
250 g lean pork

Seasoning

1/4 tsp sea salt

Method

1 Peel beetroot and tomatoes. Cut into chunks. Set aside. Slice lean pork thickly. Blanch in boiling water and drain.
2 Put all ingredients into a pot. Add 1 litre of cold water. Bring to the boil over high heat. Then turn to medium-low heat and cook for 1 hour. Season with sea salt. Serve.

Indications and contraindications

This soup looks gorgeous and tastes divine. It is suitable for those with Blood-Asthenia, dull yellow complexion, low spirits, or low sex drive. However, those with low blood pressure should avoid.

Suggested dosage

Serve once or twice each week.

Beef basil soup

(makes 2 servings) Ref.p.026

Ingredients

50 g fresh basil
3 slices ginger
250 g beef

Seasoning

1/4 tsp sea salt

Method

1 Rinse the basil and pick off the leaves. Discard the stems. Rinse the beef and slice thinly.
2 Boil 600 ml of water in a pot. Put in beef and ginger. Boil for 10 minutes. Put in the basil leaves. Boil for 3 minutes. Season with sea salt. Serve.

Indications and contraindications

This soup is rich and tasty, and is very easy to make. This is suitable for those with anaemia, headache, dysmenorrhoea, low spirits, insufficient Qi, general weakness or those having problem conceiving. However, those with Qi-Asthenia and Blood-Dryness, and those with recurrent skin problem should avoid.

Suggested dosage

Serve once or twice each week.

Squab soup with cordyceps flowers and goji berries

(makes 2 to 3 servings) Ref.p.028

Ingredients

6 g cordyceps flowers
5 g dried goji berries
3 slices ginger
6 red dates
1 squab

Seasoning

1/4 tsp sea salt

Method

1 Rinse and soak cordyceps flowers and goji berries in water. Drain and set aside. De-seed the red dates. Dress the squab and rinse well. Chop into pieces and blanch in boiling water. Drain.
2 Put all ingredients into a pot. Add 1 litre of cold water. Boil over medium-low heat for 1 hour. Season with sea salt. Serve.

Indications and contraindications

This soup is flavourful and sweet, while nourishing the body. It is suitable for those with Kidney-Asthenia, general weakness, low energy, insufficient Qi (vital energy) and Jing (essence of life), restlessness, erectile dysfunction and low sex drive. However, those with cold or flu with fever should not consume.

Suggested dosage

Serve once every three days.

Silkie chicken soup with peanuts and goji berries

(makes 2 to 3 servings) Ref.p.030

Ingredients

30 g peanuts
6 g dried goji berries
6 red dates
3 slices ginger
1/2 silkie chicken

Seasoning

1/4 tsp sea salt

Method

1   Soak and rinse peanuts and goji berries in water. Drain and set aside. De-seed the red dates. Dress the silkie chicken and rinse well. Chop into big pieces. Blanch in boiling water. Drain.

2   Put all ingredients into a pot. Add 1 litre of cold water. Bring to the boil. Cook over medium-low heat for 1 hour. Season with sea salt. Serve.

Indications and contraindications

This soup is tasty and nourishing without causing Dryness. It is suitable for those with general weakness, Blood-Asthenia, poor memory, weak Spleen and Stomach, or dysmenorrhoea. However, those having cold or flu with fever and those with overwhelming Yang energy in the Liver should avoid.

Suggested dosage

Serve once or twice each week.

Double-steamed chicken soup with Huai Shan and sea cucumber

(makes 2 to 3 servings)   Ref.p.032

Ingredients

30 g Huai Shan (dried yam)
5 g dried goji berries
200 g rehydrated sea cucumber
4 red dates
3 slices ginger
1/2 free-range chicken

Seasoning

1/4 tsp sea salt

Method

1   Rinse and soak Huai Shan and goji berries in water. Drain and set aside. Cut sea cucumber into pieces. Blanch chicken and sea cucumber in boiling water. Drain and set aside. De-seed the red dates.

2   Put all ingredients into a double-steaming pot. Pour in 600 ml of boiling water. Double-steam in a water bath for 2 hours. Season with sea salt. Serve.

Indications and contraindications

This soup is tasty and fragrant. It nourishes the Yin and strengthen the body. It is suitable for those with Liver- and Kidney-Asthenia, Spleen-Asthenia with loose stool, malnutrition, dysmenorrhoea, nocturnal emission or erectile dysfunction. However, those with Dampness-Heat body type and those not having completely recovered from cold or flu should not consume.

Suggested dosage

Serve once every three days.

Oyster and radish soup

(makes 2 to 3 servings) Ref.p.035

Ingredients

150 g frozen oysters (shelled, medium-sized)
1 white radish
1 tbsp shredded ginger
2 sprigs basil (leaves only)
500 ml stock

Seasoning

1/2 tsp sea salt
ground white pepper

Method

1   Thaw the oysters. Blanch in boiling water briefly. Drain and set aside. Peel the radish and shred finely.
2   Stir-fry the shredded ginger in a pot with a little oil until fragrant. Pour in the stock and shredded radish. Boil over high heat for 15 minutes. Put in the oysters and basil leaves. Cook for 5 minutes. Season with sea salt and ground white pepper. Serve.

Indications and contraindications

This soup is flavourful and delicious. It nourishes and strengthen the body. It is suitable for those with Heat in the Stomach meridian, constipation, excessive sweating without exercising, or low semen volume. Those suffering from gout should not consume.

Suggested dosage

Serve once or twice each week.

Quail soup with Himematsutake and almonds

(makes 2 to 3 servings) Ref.p.038

Ingredients

10 g dried Himematsutake mushrooms
10 g almonds
3 slices ginger
6 red dates
1 quail

Seasoning

1/2 tsp sea salt

Method

1 Soak Himematsutake mushrooms in water until soft. Drain and set aside. Rinse the almonds. De-seed the red dates. Dress the quail and rinse well. Chop into pieces and blanch in boiling water. Drain.
2 Put all ingredients into a pot. Add 1.5 litres of water. Bring to the boil over high heat. Turn to medium-low heat and cook for 30 minutes. Season with sea salt. Serve.

Indications and contraindications

This soup is fragrant and tasty. It nourishes and benefits the body in general. It is suitable for those with insufficient Qi (vital energy) and Blood, cough and shortness of breath with much phlegm, general weakness with frequent sickness, or low energy and low spirits. However, those not having completely recovered from cold or flu and those allergic to mushrooms should avoid.

Suggested dosage

Serve once or twice each week.

Egg drop soup with lily bulbs

(makes 2 servings) Ref.p.040

Ingredients

30 g dried lily bulbs
1 slice cooked ham
2 tsp shredded ginger
1 egg
600 ml stock

Seasoning

1/4 tsp sea salt

Method

1 Rinse and soak the lily bulbs in water. Drain and set aside. Finely shred the ham. Whisk the egg.
2 Boil the stock in a pot. Add lily bulbs and ginger. Boil for 20 minutes. Stir in the whisked egg. Add shredded ham and season with salt. Serve.

Indications and contraindications

This soup is tasty and very easy to make. It is suitable for those with restlessness, insomnia, poor sleep quality, poor appetite or diminishing brain power. Generally speaking, everyone may consume.

Suggested dosage

Serve once every three days.

Cream of pumpkin soup

(makes 2 to 3 servings) Ref.p.043

Ingredients

200 g pumpkin
1/2 onion
500 ml stock
200 ml whipping cream
25 g butter
15 g flour
2 cloves garlic
1 tbsp finely chopped parsley (optional)

Seasoning

1/2 tsp sea salt

Method

1 Peel the pumpkin. Rinse and dice it. Set aside. Peel and dice the onion. Crush the garlic cloves with the flat side of a knife.

2 Melt butter in a pot over low heat. Put in onion and garlic. Stir-fry until fragrant. Turn to high heat and add diced pumpkin. Toss until pumpkin is tender and juices released. Sprinkle with flour and toss to pick up the juices. Pour in stock and stir well. Turn off the heat.

3 Blend the mixture with a hand-blender (or pour it into a blender, puree and pour it back in the pot). Turn on low heat and bring to the boil. Turn off the heat. Stir in whipping cream and season with sea salt. Sprinkle with chopped parsley on top. Serve.

Indications and contraindications

This soup is creamy and tasty. It is good for all ages. It is especially suitable for those with compromised digestive function, poor appetite, physical and mental exhaustion, or low sperm count. However, those with accumulated Dampness in the body should consume in moderation.

Suggested dosage

Serve once or twice each week.

Pre-pregnancy soups for men

Walnut cockerel soup

(makes 2 to 3 servings) Ref.p.046

Ingredients

30 g shelled walnuts
4 black dates
3 slices ginger
1 cockerel

Seasoning

1/2 tsp sea salt

Method

1 Rinse the walnuts and black dates. Dress the chicken and rinse well. Chop into large chunks. Blanch in boiling water. Drain.
2 Put all ingredients into a pot. Add 1.2 litres of water. Bring to the boil over high heat. Turn to medium-low heat and cook for 90 minutes. Season with sea salt. Serve.

Indications and contraindications

This soup is tasty and full of umami. It nourishes and strengthens the body. It is suitable for those prone to coughs and shortness of breath, those with general physical weakness, dizziness and palpitations, frequent urinations, involuntary ejaculations, or thin and watery semen. Those having cold or flu with fever, and those having recurring skin problems should not consume cockerel.

Suggested dosage

Serve once or twice each week.

Pre-pregnancy soups for men

Dried oyster soup with gingkoes, beancurd skin and shiitake mushrooms

(makes 2 to 3 servings) Ref.p.048

Ingredients

15 gingkoes
1/2 sheet beancurd skin
4 dried shiitake mushrooms
50 g dried oysters
1 piece dried tangerine peel

Seasoning

1/2 tsp sea salt

Method

1 Shell and core the gingkoes. Soak dried tangerine peel in water till soft. Soak dried shiitake mushrooms in water till soft. Cut off the stems and set aside. Rinse the dried oysters well.
2 Put all ingredients into a pot. Add 1.5 litres of water. Bring to the boil over high heat. Turn to medium-low heat and cook for 30 minutes. Season with sea salt. Serve.

Indications and contraindications

This soup is rich and flavourful. It nourishes the body without causing Dryness or Fire. It is suitable for those with weak lungs who are prone to coughs and shortness of breath; those with osteoporosis, involuntary ejaculation, leucorrhoea, and deteriorating energy level. However, those with Spleen- or Stomach-Asthenia with Coldness should consume in moderation.

Suggested dosage

Serve once or twice each week.

Egg drop soup with laver and tomatoes

(makes 2 to 3 servings) Ref.p.050

Ingredients

5 g laver
2 tomatoes
2 eggs
1 tbsp finely chopped spring onion
2 tsp finely shredded ginger

Seasoning

1/2 tsp sea salt

Method

1 Soak and rinse laver. Peel tomatoes and cut into chunks. Whisk the eggs.
2 Boil 600 ml of water in a pot. Put in laver, tomatoes and shredded ginger. Boil for 15 minutes. Season with sea salt. Stir in the whisked egg and keep stirring until it boils again. Sprinkle with spring onion on top. Serve.

Indications and contraindications

This soup is tasty and easy to make. It is suitable for those with anaemia, poor digestion, difficulty urinating, overweight, oedema or infertility. Generally speaking, everyone can consume. Only those with hyperthyroidism should not consume laver.

Suggested dosage

Serve once or twice each week.

Dried mussel soup with soybean sprouts

(makes 2 to 3 servings) Ref.p.053

Ingredients

200 g soybean sprouts
15 g small dried shiitake mushrooms
50 g dried mussels
1 carrot
3 slices ginger

Seasoning

1/2 tsp sea salt

Method

1 Cut off both ends of each soybean sprout. Fry them in a dry wok. Set aside. Rinse and soak shiitake mushrooms in water till soft. Cut off the stems. Set aside. Rinse and soak dried mussels in water. Drain and blanch in boiling water. Drain and set aside. Peel and cut carrot into chunks.
2 Boil 1.5 litres of water in a pot. Put in all ingredients and bring to the boil. Turn to medium-high heat. Cook for 1 hour. Season with sea salt. Serve.

Indications and contraindications

This soup is flavourful and delicious. It nourishes and beautifies the skin. It is suitable for those with spots and warts on the skin, those with Kidney-Asthenia accompanied by oedema, high blood pressure, high triglyceride levels, involuntary ejaculation, leucorrhoea, or low energy level. Generally speaking, everyone may consume. Only those suffering from gout should avoid.

Suggested dosage

Serve once or twice each week.

Garlic clam soup

(makes 2 servings) Ref.p.056

Ingredients

6 to 8 cloves garlic
5 to 6 sprigs yellow chives
250 g live large clams
1 tsp finely chopped chilli
500 ml stock

Seasoning

sea salt
ground white pepper
rice wine

Method

1 Peel the garlic and crush it gently. Rinse the yellow chives and cut into short lengths. Soak the clams in salted water for six hours to let them spit out any sand.
2 Boil the stock and garlic in a pot. Put in the clams. Cook until they open. Add rice wine, sea salt and white pepper. Sprinkle with yellow chives and red chilli. Bring to the boil again. Serve.

Indications and contraindications

This soup is tasty and full of umami. It is also nourishing and good for the body. It is especially suitable for those with poor appetite, physical exhaustion, beriberi, oedema, low semen volume, diabetes, high triglyceride levels, or high blood pressure. However, those suffering from gout should not consume shellfish.

Suggested dosage

Serve once or twice each week.

Cockerel soup with shrimps and goji berries

(makes 2 servings) Ref.p.058

Ingredients

100 g shrimps
5 g dried goji berries
3 slices ginger
6 red dates
1/2 cockerel

Seasoning

1/4 tsp sea salt

Method

1 Shell and devein the shrimps. Soak and rinse goji berries in water. De-seed the red dates. Dress the cockerel and rinse well. Chop into pieces. Blanch in boiling water. Drain and set aside.
2 Put chicken, red dates and ginger into a pot. Add 1.2 litres of water. Bring to the boil over high heat. Turn to medium heat and cook for 30 minutes. Add goji berries and shrimps. Boil for 5 minutes. Season with sea salt. Serve.

Indications and contraindications

This soup in tasty and it nourishes the body. It is suitable for those with general weakness due to prolonged exhaustion, Kidney-Asthenia accompanied by premature ejaculation or erectile dysfunction, abnormally heavy menstrual bleeding, or frequent urinations. Those with gout or those not completely recovering from cold or flu should not consume.

Suggested dosage

Serve once or twice each week.

Cream of purple yam soup with sweet corn

(makes 2 servings) Ref.p.060

Ingredients

100 g purple yam
20 g frozen sweet corn kernels
500 ml milk
2 egg whites
2 tbsp caltrop starch slurry (one part caltrop starch mixed with one part water)

Seasoning

1/2 tsp sea salt
ground white pepper

Method

1 Peel the purple yam. Rinse and cut into chunks. Put purple yam and 250 ml of milk into a blender. Puree till fine.

2 Pour the purple yam puree into a pot. Add sweet corn kernels and the remaining milk. Cook over low heat until it boils. Stir in caltrop starch slurry and season with sea salt and white pepper. Stir continuously while heating until it thickens and boils. Stir in egg white. Bring to the boil and serve. (Don't be alarmed if the egg white turns blue after cooked. It's because it has picked up the anthocyanin from the purple yam.)

Indications and contraindications

This soup is creamy and tasty. It nourishes the Yin and alleviates Asthenia. It is suitable for those under immense stress from work, those with general weakness due to prolonged sickness, poor Qi (vital energy) and Blood flow, malnutrition, or infertility. Generally speaking, everyone may consume.

Suggested dosage

Serve once or twice each week.

Abalone soup with fresh Shi Hu and goji berries

(makes 2 servings) Ref.p.063

Ingredients

15 g fresh Shi Hu (dendrobium stems)
6 g dried goji berries
6 live small abalones
150 g lean pork

Seasoning

1/4 tsp sea salt

Method

1 Rinse the Shi Hu and cut into short lengths. Rinse and soak goji berries in water. Remove the innards from the abalones. Rinse well and set aside. Slice the pork thickly. Blanch in boiling water. Drain and set aside.
2 Put all ingredients into a pot. Add 1.2 litres of water. Bring to the boil and turn to medium-low heat. Cook for 1 hour. Season with sea salt. Serve.

Indications and contraindications

This soup is flavourful and sweet. It nourishes the body without causing Dryness. It is suitable for those with deteriorating eyesight, dizziness, tinnitus, Jing-Asthenia due to prolonged exhaustion, Blood-Asthenia with yellow dull complexion, soreness and weakness in the knees and lower back, diabetes, or fatty liver. Gout patients should not consume too much shellfish.

Suggested dosage

Serve once or twice each week.

Double-steamed quail soup with fish maw and conch

(makes 2 to 3 servings) Ref.p.066

Ingredients

150 g rehydrated fish maw
3 frozen conches
3 slices ginger
6 red dates
2 quails

Seasoning

1/2 tsp sea salt

Method

1 Rinse the fish maw and conches. Blanch in boiling water. Drain and set aside. De-seed the red dates. Dress the quails and rinse well. Blanch in boiling water. Drain.
2 Put all ingredients into a double-steaming pot. Pour in boiling water to cover all ingredients. Cover the lid and double-steam in a boiling water bath for 3 hours. Season with salt. Serve both the soup and the solid ingredients.

Indications and contraindications

This soup is rich and flavourful. It nourishes and strengthens the body. It is suitable for those with physical weakness, anaemia, dizziness, low spirits, or low sex drive. Those suffering from cold or flu with fever and those with poor digestion should not consume.

Suggested dosage

Serve once or twice each week.

Black sesame and walnut milk

(makes 2 servings) Ref.p.069

Ingredients

2 tbsp ground black sesame
2 tbsp ground walnut
500 ml milk
2 tbsp cornstarch slurry (1 part cornstarch mixed with 1 part water)

Seasoning

Organic agave syrup

Method

1 Put the ground black sesame and ground walnut into a bowl. Stir in a little milk and mix into a thick paste without any dry patch.
2 Heat the remaining milk in a pot. Add the black sesame and walnut paste from step 1. Bring to the boil and stir in cornstarch slurry. Keep stirring while bringing to the boil. Stir in the agave syrup. Serve.

Indications and contraindication

This drink is creamy and nutty. It is good for all ages. It is especially suitable for those with habitual constipation, coarse skin texture, poor blood circulation, deteriorating memory, or infertility. Those with Spleen-Asthenia suffering from loose stools or diarrhoea should not consume.

Suggested dosage

Serve two to three times each week.

Walnut and hazelnut milk

(makes 2 servings)

Ingredients

30 g shelled walnuts
30 g shelled hazelnuts
500 ml milk
2 tbsp cornstarch slurry (1 part cornstarch mixed with 1 part water)

Seasoning

1 tbsp sugar

Method

1 Grind the walnuts and hazelnuts till fine in a grinder. Add cold drinking water and mix into a smooth paste.
2 Heat the milk in a pot. Stir in the walnut and hazelnut paste from step 1. Add sugar and bring to a gentle simmer. Stir in cornstarch slurry. Keep stirring until it boils again. Serve.

Indications and contraindications

This drink is sweet and tasty. It is good for all ages. It is especially suitable for those with insufficient Qi (vital energy) in the Kidney meridian, deteriorating brain power, soreness and weakness in the knees and lower back, weak Lungs with proneness to coughs and shortness of breath, erectile dysfunction, involuntary ejaculation, or frequent urinations. However, those suffering from diarrhoea or loose stools, and those with Yin-Asthenia and overwhelming Fire should not consume.

Suggested dosage

Serve two to three times each week.

Black soymilk with oats

(makes 2 servings) Ref.p.074

Ingredients

30 g instant oats
500 ml black soymilk
1 to 2 red dates (de-seeded and thinly sliced)

Seasoning

raw cane sugar

Method

Heat the black soymilk. Add oats and sliced red dates. Cook for 5 minutes. Stir in sugar and cook until it dissolves. Serve.

Indications and contraindications

This drink is sweet and tasty. It nourishes and beautifies the skin. It is especially suitable for those with high triglyceride levels, high blood pressure, high blood cholesterol level, osteoporosis, or infertility. Gout patients should avoid.

Suggested dosage

Serve two to three times each week.

Oatmeal with chia seeds

(makes 2 servings) Ref.p.076

Ingredients

1 tbsp chia seeds
30 g rolled oats
500 ml milk

Seasoning

raw cane sugar

Method

Heat up the milk and add rolled oats and chia seeds. Cook over low heat for 5 minutes. Add sugar and mix well. Serve.

Indications and contraindications

This porridge is creamy and tasty. It is good for all ages. It is especially recommended to those with high blood pressure, high blood glucose level and high triglyceride levels; or those with overweight problem, oedema, extreme mood swings, osteoporosis or infertility. Generally speaking, everyone may consume.

Suggested dosage

Serve two to three times each week.

Du Zhong tea

(makes 2 servings) Ref.p.078

Ingredients

2 tbsp dried young shoots of Du Zhong leaves

Method

Put the Du Zhong leaves into a teapot. Pour in boiling water to rinse the leaves once. Drain and refill it with boiling water. Cover the lid and leave it for 5 minutes. Serve.

Indications and contraindications

This tea is mildly bitter, with a fragrant aroma. It is suitable for those with Liver- or Kidney-Asthenia, dizziness, insufficient Qi (vital energy) in the Kidney meridian, erectile dysfunction, or involuntary ejaculation. Frequent consumption of Du Zhong leaves would promote the metabolism of sugar in the body. Therefore, diabetics should not consume this drink.

Suggested dosage

Serve two to three times each week.

Wheat germ tea

(makes 2 servings) Ref.p.080

Ingredients

2 tbsp ground wheat germ

Seasoning

honey

Method:

Put ground wheat germ into a serving mug. Add boiling water and stir to mix well. Season with honey. Mix again. Serve.

Indications and contraindications

This tea is fragrant and is good for all ages. It is especially recommended to those with high blood glucose level, high triglyceride levels, high blood cholesterol, and those with low energy level. Generally speaking, everyone may consume.

Suggested dosage

Serve two to three times each week.

Red bean and black glutinous rice tea

(makes 2 servings) Ref.p.082

Ingredients

30 g red beans
30 g black glutinous rice

Seasoning

rock sugar

Method

1 Soak red beans and black glutinous rice in water for 2 hours. Drain.
2 Boil 800 ml of water in a pot. Add red beans and black glutinous rice. Cook for 40 minutes. Add rock sugar and cook until it dissolves.
3 Serve the tea along with some red beans and black glutinous rice.

Indications and contraindications

This tea is sweet and aromatic. It is good for all ages. It is especially recommended to those with overweight problem, oedema, physical and mental exhaustion, anaemia, pale complexion, poor digestion, or nervous prostration. Generally speaking, everyone may consume.

Suggested dosage

Serve two to three times each week.

Apple cinnamon tea

(makes 2 servings) Ref.p.085

Ingredients

2 apples
2 cinnamon sticks (or 1/2 tsp ground cinnamon)

Seasoning

organic agave syrup

Method

1 Soak apples briefly in hot water. Then rinse off the fertilizers and wax on the skin. Core them and cut into chunks.
2 Boil 400 ml of water in a pot. Put in apples and cinnamon. Cook for 15 minutes. Stir in the agave syrup. Mix well and serve.

Indications and contraindications

This tea is aromatic and rich. It warms the Stomach and expels Coldness. It is suitable for those with Asthenia-Coldness in the Spleen and Stomach meridians; cold limbs, indigestion, poor blood circulation, diabetes, or infertility. Those with accumulated Heat and overwhelming Fire, and pregnant women should avoid.

Suggested dosage

Serve two to three times each week.

Vegetarian dices in ginger juice and oyster sauce

(makes 2 to 3 servings) Ref.p.089

Ingredients

6 large semi-dried oysters
1 small segment carrot
3 asparaguses
1 piece five-spice dried tofu
2 dried shiitake mushrooms
2 tbsp ginger juice

Seasoning

1 tbsp light soy sauce
2 tsp oyster sauce
1/2 tsp sea salt
rice wine

Method

1   Soak and rinse the semi-dried oysters in water. Steam for 30 minutes. Dice coarsely and set aside. Peel carrot and dice it coarsely. Rinse the asparaguses and dice them coarsely. Dice the five-spice dried tofu coarsely. Soak shiitake mushrooms in water till soft. Cut off the stems. Dice them.

2   Heat some oil in a wok. Fry the diced semi-dried oysters until fragrant. Put in all remaining ingredients. Toss well. Sizzle with rice wine. Add ginger juice and all remaining seasoning. Sprinkle with some water. Cook for 10 minutes until the sauce reduces. Serve.

Indications and contraindications

This dish is rich and tasty. It is good for all ages. It is especially recommended to those with Yin-Asthenia, restlessness and fever, insufficient Qi (vital energy) and Blood, mental exhaustion, anaemia, or infertility. Those with Asthenia-Coldness in the Spleen and Stomach meridians, and those suffering from gout should not consume.

Suggested dosage

Serve once or twice each week.

Broccoli in dried scallop glaze

(makes 2 to 3 servings) Ref.p.092

Ingredients

4 dried scallops (soaked in water till soft, steamed for 15 minutes, broken into fine shreds)
1 head broccoli
1 tsp grated ginger
100 ml stock
2 tbsp caltrop starch slurry (1 part caltrop starch mixed with 1 part water)

Seasoning

1/2 tsp sea salt
1 tbsp oyster sauce

Method

1 Rinse the broccoli. Cut into florets. Blanch in boiling water with a dash of oil and a pinch of salt until cooked through. Drain and arrange on a serving plate.
2 Heat a little oil in a wok. Stir-fry grated ginger until fragrant. Add the steamed dried scallops, stock and seasoning. Bring to the boil. Stir in caltrop starch slurry and cook until it thickens. Pour the mixture over the bed of broccoli on the serving plate. Serve.

Indications and contraindications

This dish is sweet and tasty. It nourishes the Yin and alleviates Asthenia. It is suitable for those with high blood pressure, high triglyceride and blood glucose levels; those with diabetes, overweight problem, poor appetite, or difficulty conceiving. Generally speaking, everyone may consume.

Suggested dosage

Serve once or twice each week.

Stir-fried diced chicken with bell pepper and walnuts

(makes 2 to 3 servings) Ref.p.095

Ingredients

150 g tri-colour bell peppers
100 g candied walnuts
200 g boneless chicken thigh
2 tsp grated ginger

Seasoning

1/2 tsp sea salt
rice wine

Marinade

2 tsp light soy sauce
2 tsp rice wine
1/4 tsp salt
caltrop starch
ground white pepper

Method

1 Rinse the bell peppers. De-seed and dice them. Set aside. Rinse the chicken thigh. Dice it and add marinade. Mix well. Leave it for 30 minutes.

2 Heat some oil in a wok. Stir-fry grated ginger until fragrant. Put in the diced chicken and toss well. Add bell peppers and toss again. Sizzle with rice wine and add sea salt. Sprinkle with candied walnuts. Toss well and serve.

Indications and contraindications

This dish is sweet and delicious, while nourishing and strengthening the body. It is suitable for those with skinny build due to prolonged exhaustion, poor appetite, deteriorating brain power, soreness and weakness in the knees and lower back, erectile dysfunction, or involuntary ejaculation. Generally speaking, everyone may consume.

Suggested dosage

Serve once or twice each week.

Scrambled egg with Chinese chives and shrimps

(makes 2 to 3 servings) Ref.p.098

Ingredients

60 g Chinese chives
100 g shelled shrimps
2 eggs

Seasoning

1/2 tsp sea salt
1 tbsp oil
ground white pepper

Method

1   Trim off the yellowish ends of the Chinese chives. Rinse and finely chop them. Set aside. Devein the shrimps. Rinse and drain well.
2   Whisk the eggs. Add seasoning and whisk again.
3   Heat 2 tbsp of oil in a wok. Stir-fry the shrimps until cooked through and lightly browned. Add Chinese chives and toss briefly. Pour in the eggs and stir until the eggs are just cooked.

Indications and contraindications

This dish is aromatic and tasty. It nourishes and strengthens the body. It is suitable for those with low spirits, mental and physical exhaustion, soreness and weakness in the knees and lower back, Kidney-Asthenia with erectile dysfunction, or menstrual disorder. Generally speaking, everyone may consume. Yet, those with gout should consume in moderation.

Suggested dosage

Serve once or twice each week.

Oyster omelette, Chuichow style

(makes 2 to 3 servings) Ref.p.100

Ingredients

250 g baby oysters
2 duck eggs
1 tbsp finely chopped spring onion
1 sprig coriander (finely chopped)

Batter

2 tbsp sweet potato starch
1 tbsp cornstarch
1/2 cup water

Seasoning

1 tsp salt
ground white pepper

Method

1   Rub the baby oysters in caltrop starch and salt. Rinse well and drain. Set aside. Whisk the duck eggs.
2   Make the batter by mixing the ingredients together. Add seasoning and mix again. Add the baby oysters from step 1.
3   Heat 4 tbsp of oil in a wok. Pour in the batter and fry briefly. Then pour in the duck eggs. Fry until both sides golden. Sprinkle with coriander and ground white pepper. Serve.

Indications and contraindications

This dish is crispy on the outside and juicy on the inside. It is delicious and appetizing. It is especially recommended to those with Spleen- or Stomach-Asthenia, anaemia, low spirits, or osteoporosis. However, those with poor Qi (vital energy) flow, fatty liver, or gout should consume in moderation.

Suggested dosage

Serve once or twice each week.

Korean braised beef ribs with radish

(makes 3 servings) Ref.p.103

Ingredients:

400 g beef ribs
200 g white radish
1 red chilli
2 cloves garlic
3 slices ginger
1 tbsp finely chopped coriander

Seasoning

1/2 bottle Korean marinade for barbecue beef
1/2 tsp salt
1 tbsp rice wine

Method

1 Blanch the beef ribs in boiling water. Drain and set aside. Peel white radish and cut into chunks. De-seed the chilli and dice it.

2 Fry beef ribs in a wok with a little oil until both sides golden. Add all other ingredients (except chopped coriander) and seasoning. Top with water to cover all ingredients. Bring to the boil over high heat. Turn to medium heat and cook for 1 hour until the sauce reduces and thickens. Sprinkle with coriander. Serve.

Indications and contraindications

This dish is rich and flavourful. It nourishes and strengthens the body. It is especially recommended to those with Qi- and Blood-Asthenia, malnutrition, anaemia, dull yellow complexion, soreness in the joint, or low sex drive. Generally speaking, everyone may consume. Yet, those with recurring skin problem should consume in moderation.

Suggested dosage

Serve once or twice each week.

Braised squab in fennel sauce

(makes 2 servings) Ref.p.108

Ingredients

1 squab
1 fennel bulb
5 g dried goji berries
2 slices ginger

Seasoning

2 tbsp spiced soy marinade
1 tbsp light soy sauce
1/4 tsp sea salt
4 cubes white rock sugar

Method

1 Rinse the fennel bulb and thinly slice it. Set aside. Rinse the goji berries. Dress the squab and rinse well.

2 Heat some oil in a wok. Stir-fry ginger and fennel until fragrant. Put in the squab to sear on all sides briefly. Add goji berries, seasoning and 200 ml of water. Bring to the boil and turn to low heat. Cook for 15 minutes. Remove the squab from the sauce. Chop into pieces and arrange on a serving plate. Heat the sauce up and drizzle over the squab. Serve.

Indications and contraindications

This dish is tasty and nourishing. It is suitable for those with poor digestion due to weak Spleen and Stomach, infertility due to Coldness in the womb, erectile dysfunction, involuntary ejaculation or infertility. However, those with Yin-Asthenia accompanied by overwhelming Fire and pregnant women should not consume.

Suggested dosage

Serve once or twice each week.

Braised fish maw with leafy greens

(makes 2 to 3 servings) Ref.p.106

Ingredients

6 small fish maws (rehydrated)
6 sprigs Shanghainese baby Bok Choy
3 slices ginger
3 sprigs spring onion (white parts only)

Seasoning

2 tbsp oyster sauce
150 ml chicken stock
rice wine
1 tbsp caltrop starch slurry (1 part caltrop starch mixed with one part water)

Method

1 Blanch the fish maws in boiling water with ginger and white parts of spring onion. Drain and set aside.
2 Rinse the Shanghainese baby Bok Choy. Cut each in half lengthwise. Blanch in boiling water until cooked through. Arrange neatly on a serving plate.
3 Boil chicken stock in a pot. Put in the fish maws and cook for 10 minutes or until desired consistency. Arrange fish maws on the serving plate over the baby Bok Choy. Add oyster sauce and rice wine to the chicken stock. Bring to the boil and stir in caltrop starch slurry. Cook until it thickens. Drizzle the glaze over the fish maw and baby Bok Choy. Serve.

Indications and contraindications

This dish is taste and nourishing. It is especially suitable for those with general weakness, low spirits, physical and mental exhaustion, soreness in the lower back, or low sex drive. Generally speaking, everyone may consume. Yet, those having cold or flu with fever should avoid.

Suggested dosage

Serve once or twice each week.

Three-cup chicken

(makes 2 to 3 servings) Ref.p.111

Ingredients

30 g fresh basil
1 spring chicken
8 cloves garlic
2 red chillies
20 g ginger

Seasoning

1 small cup sesame oil
1 small cup thicken soy paste
1 small cup rice wine
2 tsp sugar

Method

1 Dress and rinse the chicken. Chop into pieces and set aside. Pick the leaves off the basil and discard the stems. Rinse the basil leaves and set aside. Slice the ginger. Peel the garlic. Slice the red chilli.
2 Heat sesame oil in a wok. Fry the chicken until golden. Set aside.
3 In the same wok, fry the ginger, garlic and red chilli until fragrant. Put the chicken back in and toss until fragrant again. Sizzle with rice wine and add the remaining seasoning. Bring to the boil and turn to medium-low heat. Cook for 10 minutes until the sauce reduces. Sprinkle with basil leaves. Toss briefly. Serve.

Indications and contraindications

This dish is rich and aromatic. It nourishes and strengthens the body. It is especially suitable for those with Asthenia-Coldness in the Spleen and Stomach meridians, poor appetite, low spirits, weakness in the lower back and knees, and those having difficulty conceiving. Generally speaking, everyone may consume. Only those suffering from cold or flu with fever and those with Yin-Asthenia and overwhelming Fire should consume in moderation.

Suggested dosage

Serve once or twice each week.

Steamed mussels in garlic black bean sauce

(makes 2 to 3 servings) Ref.p.114

Ingredients

10 frozen mussels (shelled)
2 tsp fermented black beans
2 tsp grated garlic
2 tsp grated ginger
2 tsp finely chopped red chilli
2 tsp finely chopped coriander

Seasoning

cooked oil
light soy sauce

Method

1 Thaw the mussels. Rinse and drain well. Arrange on a steaming plate. Set aside. Finely chop the fermented black beans. Mix together black beans, garlic, ginger and red chilli. Mix well. Spread the mixture evenly on each mussel. Steam over high heat for 7 minutes.
2 Sprinkle with coriander. Sizzle with smoking hot oil. Drizzle with light soy sauce. Serve.

Indications and contraindications

This dish is flavourful and tasty. It is good for all ages. It is suitable for those with skinny build due to prolonged exhaustion, anaemia, dizziness, night sweat, erectile dysfunction, involuntary ejaculation, leucorrhoea, and infertility due to Coldness in the womb. Generally speaking, everyone may consume. Yet, those with gout should consume in moderation.

Suggested dosage

Serve once or twice each week.

Blanched dumplings with pork and Chinses chive filling

(makes 2 to 3 servings) Ref.p.120

Ingredients

100 g Chinese chives
300 g ground pork
1 egg white
1 tsp grated ginger
200 g dumpling skin

Seasoning

1/2 tsp sea salt
1 tsp chicken bouillon powder
1/2 tsp sugar
1 tbsp rice wine
1 tbsp cornstarch

Dipping sauce

2 tsp finely shredded ginger
sesame oil
dumpling vinegar

Method

1 Rinse the chives. Trim off the old leaves. Finely chop them. Add a few drops of oil and stir well to stop the chives from giving juices. In a small bowl, mix together, ground pork and chives. Stir in egg white and seasoning. Stir until sticky.
2 Wrap some filling in each dumpling skin. Seal the seam and blanch in vigorously boiling water. Cook until the dumplings float. Add 1/2 bowl of cold water. Bring to the boil and cook until the dumplings float again. Drain and serve with the dipping sauce on the side.

Indications and contraindications

This is a delicious staples that is good for all ages. It is especially suitable for those with Yang-Asthenia, cold limbs, soreness and weakness in the lower back and knees, involuntary ejaculation, and Coldness in the womb. Generally speaking, everyone may consume. Yet, those with Yin-Asthenia and overwhelming Fire should avoid.

Suggested dosage

Serve once or twice each week.

Fusilli with Brussels sprouts in avocado beef sauce

(makes 2 servings) Ref.p.117

Ingredients

1 avocado
2 tbsp extra-virgin avocado oil
6 Brussels sprouts
4 button mushrooms
50 g ground beef
150 g fusilli pasta (or rotini)

Marinade

2 tsp light soy sauce
2 tsp rice wine
1 tsp cornstarch

Seasoning

sea salt
ground white pepper

Method

1 Cut the avocado in half. Remove the core and scoop out the flesh. Mash it with a fork and set aside. Rinse the Brussels sprouts. Cut each in half and set aside. Cut off the stems of the button mushrooms and slice them. Blanch Brussels sprouts and button mushrooms in boiling water briefly. Drain and set aside. Add marinade to the ground beef. Mix well.

2 Cook the fusilli in boiling water for 12 minutes (or according to the cooking instructions on the package). Drain and transfer onto a serving plate.

3 While the fusilli is cooking, make the sauce. Heat the avocado oil in a pan. Stir-fry ground beef until lightly browned. Add Brussels sprouts, sliced mushrooms, and mashed avocado. Sizzle with rice wine and add 1/2 bowl of water. Add seasoning and cook for 5 minutes. Pour the sauce over the cooked fusilli on the serving plate. Serve.

Indications and contraindications

This is a delicious staple that is good for all ages. It is especially recommended to those prone to fatigue, and those with poor appetite, low spirits, or difficulty conceiving. Generally speaking, everyone may consume.

Suggested dosage

Serve two to three times each week.

Baby oyster congee

(makes 2 servings) Ref.p.138

Ingredients

150 g baby oysters (shelled)
500 ml fish stock
2 dried shiitake mushrooms
50 g ground pork
1 tbsp finely chopped spring onion
1 tbsp finely shredded ginger
60 g rice

Seasoning

sea salt
ground white pepper

Method

1 Rub baby oysters in caltrop starch. Rinse well. Blanch in boiling water. Drain and set aside.
2 Rinse the rice and add a dash of oil and a pinch of salt. Mix well and set aside. Soak shiitake mushrooms in water until soft. Cut off the stems. Finely shred them.
3 Put rice, shiitake mushrooms, shredded ginger, fish stock and 140 ml of water (or according to your preferred consistency) into an electric rice cooker. Turn on the congee cooking cycle and let it complete it.
4 Add ground pork and baby oysters to the congee. Cook for 5 minutes. Add seasoning and sprinkled with finely chopped spring onion. Serve.

Indications and contraindications

This congee is flavourful and appetizing. It nourishes and strengthens the body. It is suitable for those with general weakness, poor appetite, mental exhaustion, insomnia, thyroid problems, or low sex drive. Generally speaking, everyone may consume. Only gout patients should avoid.

Suggested dosage

Serve once or twice each week.

Purple yam congee with dried longans and lily bulbs

(makes 3 servings) Ref.p.132

Ingredients

100 g fresh purple yam
30 g dried longans (shelled and de-seeded)
30 g dried lily bulbs
60 g rice

Seasoning

rock sugar

Method

1 Peel the purple yam and rinse well. Dice it coarsely.
2 Soak and rinse lily bulbs in water. Rinse the dried longans. Rinse the rice.
3 Put all ingredients into an electric rice cooker. Add 640 ml of water (or according to your preferred consistency). Turn on the congee cooking cycle and let it complete it. Add rock sugar and cook until it dissolves. Serve.

Indications and contraindications

This congee is creamy and tasty. It is good for all ages. It is suitable for those who feel exhausted easily, those with anaemia, insomnia, diarrhoea, uterine prolapse, involuntary ejaculation or infertility. Generally speaking, everyone may consume. Only those with Dampness-Heat body type, and those suffering from constipation should avoid.

Suggested dosage

Serve once or twice each week.

Millet congee with pumpkin seeds and goji berries

(makes 2 servings) Ref.p.134

Ingredients

20 g pumpkin seeds
6 g dried goji berries
60 g millet
1 tbsp lotus root starch

Seasoning

light brown sugar to taste

Method

1 Rinse and soak the goji berries in water. Add some cold drinking water to lotus root starch and mix into a smooth paste.
2 Rinse the millet and put it into a pot. Add pumpkin seeds and 6 bowls of water. Cook for 30 minutes. Add goji berries, light brown sugar and stir in the lotus root starch slurry. Keep stirring while heating it up. Cook until it thickens. Serve.

Indications and contraindications

This porridge is sweet and tasty. It is good for all ages. It is especially suitable for those with Spleen- or Stomach-Asthenia, anaemia, malnutrition, restlessness with dry mouth, or excessive menstrual flow. Generally speaking, everyone may consume.

Suggested dosage

Serve once or twice each week.

Coconut-scented quinoa rice with chicken and turmeric

(makes 2 to 3 servings) Ref.p.123

Ingredients

200 g chicken fillet
3 asparaguses
30 g quinoa
50 g rice
3 tsp turmeric
5 tbsp coconut milk
3 cloves garlic

Marinade

2 tsp rice wine
1/4 tsp sea salt
ground white pepper
cornstarch

Seasoning

sea salt

Method

1 Rinse the chicken fillet. Dice coarsely. Add marinade and mix well. Leave it for 30 minutes. Set aside. Rinse and dice the asparaguses. Soak quinoa in warm water for 2 hours. Rinse the rice.
2 Put quinoa and rice into a rice cooker. Add equal volume of water (or according to your preferred consistency). Turn on the rice cooker and let it complete the cooking cycle. Peel the garlic and finely chop it.
3 Heat a little oil in a wok. Stir-fry garlic until fragrant. Add chicken and asparaguses. Toss until lightly browned. Put in the quinoa rice. Sprinkle with turmeric and stir well. Add coconut milk and seasoning. Toss to mix well. Serve.

Indications and contraindications

This rice is nutty and aromatic. It nourishes and strengthens the body. It is especially suitable for those with high triglyceride levels, high blood pressure, high blood cholesterol level, overweight problem, poor appetite, deteriorating memory and premenstrual syndrome. However, those with Yin-Asthenia and overwhelming Fire, and pregnant women should not consume turmeric.

Suggested dosage

Serve once or twice each week.

Ginger fried rice with egg and scallops

(makes 2 servings) Ref.p.129

Ingredients

150 g frozen scallops
2 eggs
100 g frozen mixed vegetables
2 tsp grated ginger
2 bowls day-old rice

Marinade

2 tsp chicken bouillon powder
rice wine
cornstarch

Seasoning

1/2 tsp sea salt
ground white pepper

Method

1 Thaw the scallops. Rinse in water and add marinade. Mix well and leave them for 30 minutes. Boil a pot of water. Put the scallops in and turn off the heat. Soak them until medium-well done. Drain and set aside.
2 Thaw the mixed vegetables. Blanch in boiling water. Drain and set aside. Whisk the eggs.
3 Break the day-old rice into individual grains with your hands. Heat some oil in a wok. Pour in half of the whisked eggs. Add day-old rice, grated ginger, mixed vegetables and seasoning. Toss well. Put in scallops and the remaining whisked eggs. Toss quickly to mix well. Serve.

Indications and contraindications

This rice is fragrant and tasty. It is good for all ages. It is especially recommended to those with malnutrition, poor appetite, deteriorating brain power, or frequent urinations during sleep. Generally speaking, everyone may consume. Only gout patients should consume scallops in moderation.

Suggested dosage

Serve once or twice each week.

Korean grilled eel rice

(makes 2 servings) Ref.p.126

Ingredients

1 grilled eel
a pinch shredded young ginger pickle
a pinch shredded egg omelette
a pinch toasted white sesames
1 head butter lettuce (or iceberg lettuce)
2 bowls steamed rice

Method

1 Rinse the lettuce leaves one by one. Drain and set aside. Cut the grilled eel into pieces.
2 Lay flat a lettuce leaf. Put on some rice, grilled eel, shredded young ginger pickle, shredded egg omelette and toasted sesames. Wrap it up and eat it.

Indications and contraindications

This rice dish is tasty while nourishing and strengthening the body. It is suitable for those with skinny build due to prolonged exhaustion, anaemia, dizziness, palpitations, low spirits, rheumatism, numbness in the limbs, or low sex drive. Generally speaking, everyone may consume. Only those with Phlegm-Dampness should consume in moderation.

Suggested dosage

Serve two to three times each week.

Egg white custard with ginger juice and black soymilk

(makes 2 servings) Ref.p.141

Ingredients

2 tbsp ginger juice
250 ml black soymilk
3 egg whites

Seasoning

30 g sugar

Method

1 Add a little warm water to sugar and stir until it dissolves completely. Add egg whites, ginger juice and black soymilk. Whisk well.
2 Pour the mixture into a deep steaming dish. Skim off any foam on the surface. Cover in aluminium foil. Steam over medium heat for 12 minutes. Serve.

Indications and contraindications

This custard is velvety smooth and it nourishes the skin. It is suitable for those with infertility due to Coldness in the womb, weak and thin uterine wall, malnutrition, or other reasons for infertility. Generally speaking, everyone may consume.

Suggested dosage

Serve once or twice each week.

Poached egg sweet soup with distiller's grains

(makes 1 serving) Ref.p.144

Ingredients

100 g distiller's grains
1 egg of silkie chicken

Seasoning

2 tsp raw cane sugar

Method

1 Put distiller's grain and 1 cup (250 ml) of water into a pot. Bring to the boil. Add sugar and crack in an egg of silkie chicken. Turn off the heat and cover the lid. Leave it for 2 minutes.
2 Cook the mixture over low heat to your preferred doneness of the egg. Serve.

Indications and contraindications

This sweet soup is full of wine aroma. It nourishes the body without causing Dryness. It is suitable for those with poor Qi (vital energy) and Blood flow, poor appetite, menstrual disorders, or infertility due to Asthenia. Regular consumption helps regulate the gastrointestinal condition, while beautifying the skin and boosting the bust. Generally speaking, everyone may consume.

Suggested dosage

Serve once or twice each week.

Rose tea with He Huan Hua

(makes 1 serving) Ref.p.146

Ingredients

1 tbsp dried rose buds
1 tbsp He Huan Hua (dried albizia flowers)

Method

Put all ingredients into a teapot. Pour in boiling water and swirl to rinse them once. Drain. Fill the pot with boiling water again. Cover the lid and leave it for 7 minutes. Strain and serve.

Indications and contraindications

This tea is fragrant and floral. It nourishes the skin. It is suitable for those with nervousness, depression, insomnia, menstrual disorders, or infertility due to Coldness in the womb. However, pregnant women should not consume any floral tea.

Suggested dosage

Serve two to three times each week.

Mugwort tea with brown sugar

(makes 1 serving) Ref.p.149

Ingredients

10 g dried mugwort

Seasoning

1 tbsp light brown sugar

Method

Rinse the mugwort and put into a pot. Add 4 bowls of water. Boil for 20 minutes. Stir in light brown sugar and cook until it dissolves. Serve.

Indications and contraindications

This tea is mildly bitter in taste and it is suitable for those with menstrual disorders and infertility due to Coldness in the womb. Pregnant women may also drink it as it warms the womb, stops bleeding and secure the attachment of the foetus. It is more suitable for women to consume. Those with Yin-Asthenia accompanied by overwhelming Fire, those with intense Heat trapped in the Blood, or internal bleeding should avoid.

Suggested dosage

Serve once or twice each week.

Fu Ling and lily bulb tea

(makes 1 serving) Ref.p.152

Ingredients

dried lily bulbs
Fu Ling
2 g dried goji berries

Seasoning

sugar

Method

1 Grind the lily bulbs and Fu Ling into powder. Rinse the goji berries.
2 Boil 250 ml of water in a pot. Add goji berries, 1 tbsp of ground lily bulbs and 1 tbsp of ground Fu Ling. Bring to the boil. Stir in sugar. Serve.

Indications and contraindications

This tea is sweet and easy to make. It is good for both sexes. It clears Heat and calms the nerves. It is suitable for those with restlessness and anxiety due to Asthenia, poor sleep quality, or difficulty conceiving. Generally speaking, everyone may consume. However, those with premature ejaculation due to Asthenia-Coldness and those with Qi-Asthenia should consume with care.

Suggested dosage

Serve once or twice each week.

Haw and dried lychee sweet soup Drinks and desserts for healthy

(makes 2 servings) Ref.p.154

Ingredients

10 g dried hawthorn
10 dried lychees
6 g dried goji berries
5 red dates

Seasoning

light brown sugar

Method

1 Rinse hawthorn and goji berries. Shell the dried lychees. De-seed the red dates.
2 Put all ingredients into a pot. Add 4 bowls of water. Boil for 20 minutes. Stir in light brown sugar and cook until it dissolves. Serve both the soup and solid ingredients.

Indications and contraindications

This fruity sweet soup is sour and sweet in taste. It promotes blood cell regeneration and beautifies the skin. It is suitable for women with anaemia, menstrual disorders, menstrual pain or infertility due to Coldness in the womb. It also alleviates hernia or testicle pain among men. It is good for both sexes, but those with Yin-Asthenia accompanied by overwhelming Fire and pregnant women should avoid.

Suggested dosage

Serve once or twice each week.

Honey milk with turmeric

(makes 1 serving) Ref.p.156

Ingredients

1 tsp turmeric
250 ml milk

Seasoning

honey to taste

Method

Heat milk in a pot over low heat until steam appears, but without boiling. Stir in turmeric and mix well. Bring to a gentle simmer. Turn off the heat. Stir in honey and serve.

Indications and contraindications

This drink is golden in colour and sweet in taste. It is good for both sexes and is best served before sleep. It is suitable for those with menstrual disorders, rheumatic pain, poor blood circulation, or poor sleep quality. Regular consumption helps conceiving and prevents dementia. However, those with gallstones, Yin-Asthenia and overwhelming Fire, dry mouth and pregnant women should avoid.

Suggested dosage

Serve two to three times each week.

Avocado banana smoothie

(makes 3 servings) Ref.p.158

Ingredients

2 avocadoes
2 bananas
500 ml milk
1 tsp freshly squeezed lemon juice

Method

1 Cut avocadoes in half. Core them and scoop out the flesh. Set aside. Peel and cut bananas into chunks.
2 Put avocadoes and bananas into a blender. Puree. Add milk and blend until smooth. Squeeze in lemon juice. Serve.

Indications and contraindications

This drink is creamy and tasty. It nourishes the skin. It is suitable for those with anaemia, high blood pressure, excessive stomach acid, habitual constipation or infertility. However, those with Asthenia-Coldness in the Spleen and Stomach meridians, and those suffering from diarrhoea or loose stools should avoid.

Suggested dosage

Serve two or three times each week.

調理好體質
想要懷孕，就從吃對食物開始

**Prepare your body for pregnancy:
Start with the right diet**

作者	Author
芳姐	Cheung Pui Fong
策劃/編輯	Project Editor
譚麗琴	Catherine Tam
攝影	Photographer
細權	Leung Sai Kuen
美術設計	Art Design
羅穎思	Venus Lo

出版者

Publisher

Forms Kitchen

香港鰂魚涌英皇道1065號
東達中心1305室

Room 1305, Eastern Centre, 1065 King's Road,
Quarry Bay, Hong Kong.

電話　Tel: 2564 7511
傳真　Fax: 2565 5539
電郵　Email: info@wanlibk.com
網址　Web Site: http://www.wanlibk.com
　　　　　　 http://www.facebook.com/wanlibk

發行者

Distributor

香港聯合書刊物流有限公司
香港新界大埔汀麗路36號
中華商務印刷大廈3字樓

SUP Publishing Logistics (HK) Ltd
3/F., C&C Building, 36 Ting Lai Road,
Tai Po, N.T., Hong Kong

電話　Tel: 2150 2100
傳真　Fax: 2407 3062
電郵　Email: info@suplogistics.com.hk

承印者

Printer

中華商務彩色印刷有限公司

C & C Offset Printing Co. Ltd.

出版日期

Publishing Date

二零一八年十二月第一次印刷

First print in December 2018